W9-BWG-640

Immune

BOYD COUNTY

NOV 12 2021

PUBLIC LIBRARY

BOYD COUNTY
NOV 1 2 2021
PUBLIC LIBRARY

Immune

A Journey into the Mysterious System That Keeps You Alive

Philipp Dettmer

Random House
New York

Copyright © 2021 by Philipp Dettmer

All rights reserved.

Published in the United States by Random House, an imprint
and division of Penguin Random House LLC, New York.

RANDOM HOUSE and the HOUSE colophon are registered
trademarks of Penguin Random House LLC.

LIBRARY OF CONGRESS CATALOGING-IN-PUBLICATION DATA
Names: Dettmer, Philipp, author.
Title: Immune / Philipp Dettmer.
Description: First edition. | New York : Random House, [2021]
Identifiers: LCCN 2021010071 (print) | LCCN 2021010072 (ebook) |
ISBN 9780593241318 (hardcover) | ISBN 9780593241332 (ebook)
Subjects: LCSH: Immune system. | Immunology.
Classification: LCC QR181 .D47 2021 (print) | LCC QR181 (ebook) |
DDC 616.07/9—dc23
LC record available at https://lccn.loc.gov/2021010071
LC ebook record available at https://lccn.loc.gov/2021010072

Printed in the U.S.A.

randomhousebooks.com

9 8 7 6 5 4 3

FIRST EDITION

LSCC

Illustrations by Philip Laibacher, creative director,
Kurzgesagt—In a Nutshell
Book design by Simon M. Sullivan

BOYD COUNTY

NOV 1 2 2021

PUBLIC LIBRARY

For Cathi and Mochi

Contents

Contents

Part 3 Hostile Takeover

Part 4 Rebellion and Civil War

Introduction

IMAGINE WAKING UP TOMORROW, FEELING A BIT UNDER THE WEATHER. An annoying pain in your throat, your nose is runny, you cough a bit. All in all, not bad enough to skip work, you think, as you step into the shower, pretty annoyed about how hard your life is. While you are totally not being a whiny little baby, your immune system is not complaining. It is busy keeping you alive so you can live to whine another day. And so, while intruders roam your body, killing hundreds of thousands of your cells, your immune system is organizing complex defenses, communicating over vast distances, activating intricate defense networks, and dishing out a swift death to millions, if not billions, of enemies. All while you are standing in the shower, mildly annoyed.

But this complexity is largely hidden.

Which is a real shame because there are not many things that have such a crucial impact on the quality of your life as your immune system. It is all-embracing and all-encompassing, protecting you from bothersome nuisances like the common cold, scratches, and cuts, to life-threatening stuff from cancer and pneumonia to deadly infections like COVID-19. Your immune system is as indispensable as your heart or your lungs. And actually, it is one of the largest and most widespread organ systems throughout your body, although we don't tend to think about it in these terms.

For most of us, the immune system is a vague and cloud-like entity that follows strange and untransparent rules, and which seems to sometimes work and sometimes not. It is a bit like the weather, extremely hard to predict and subject to endless speculations and opinions, resulting in actions that feel random to us. Unfortunately many people speak about the immune system with confidence but without actually understanding it, it can be hard to know which information to trust and why. But what even *is* the immune system and how does it actually work?

Understanding the mechanisms that are keeping you alive as you read this is not just a nice exercise in intellectual curiosity; it is desperately needed knowledge. If you know how the immune system works, you can understand and appreciate vaccines and how they can save your life or the lives of your children, and approach disease and sickness with a very different mindset and far less fear. You become less susceptible to snake oil salesmen who offer wonder drugs that are entirely devoid of logic. You get a better grasp on the kinds of medication that might actually help you when you are sick. You get to know what you can do to boost your immune system. You can protect your kids from dangerous microbes while also not being too stressed-out if they get dirty playing outside. And in the very unlikely case of, say, a global pandemic, knowing what a virus does to you and how your body fights it, might help you understand what the public health experts say.

Besides all these practical and useful things, the immune system is also simply beautiful, a wonder of nature like no other. The immune system is not a mere tool to make your cough go away. It is inextricably tied into almost all other processes in your body—and while it is centrally important to keeping you alive, it is likely that it may also be the part of your body that causes your untimely death, either by failing or by being too active.

I have been fascinated and obsessed by the incredible complexity of the human immune system for the better part of a decade now. It began in university where I was studying information design and was looking for a semester project and the immune system seemed like a good idea. So I got a large pile of books about immunology and began digging in, but no matter how much I read, things just did not get less complicated. The more I learned the more impossible it seemed to simplify the immune system as every layer revealed more mechanisms, more exceptions, more complexity.

And so a project that was supposed to last the spring took over the summer and then the fall and the winter. The interactions of the parts of the immune system were too elegant and the dance they danced was too beautiful to stop learning about them. This progress fundamentally changed how I experienced and felt about my body.

When I got the flu I could no longer just complain, but had to look at my body, touch my swollen lymph nodes, and visualize what my immune cells were doing right then, which part of the network was activated, and how T

Cells killed millions of intruders to protect me. When I cut myself while being careless in the forest I felt gratitude for my Macrophages, large immune cells hunting scared bacteria and ripping them into pieces to protect the open wound from infection. After taking a bite of the wrong granola bar and suffering an allergic shock, while being rushed to the hospital, I thought about Mast Cells and IgE Antibodies and how they had almost killed me in a misguided attempt to protect me from scary foodstuffs!

When I was diagnosed with cancer at the age of thirty-two and had to undergo a couple of operations and then chemotherapy, my obsession with immunology became even more intense. One of the jobs of my immune system is to kill cancer. In this case, it had failed.

But I somehow could not be angry or too upset as I had learned how hard of a job this was for my immune cells and how hard cancer had to work to keep them in check. And as the chemo melted the cancer my thoughts again went to my immune cells invading the dying tumors and eating them up one cell after another.

Disease and sickness are scary and unsettling and I've had plenty of that in my life. But knowing how my cells, my immune system, this integral and personal part of myself, defended the entity that is me, how it fought and died and healed and restored this body I inhabit, always gave me a lot of comfort. Learning about the immune system made my life better and more interesting and it alleviated a lot of the anxiety that comes with being sick. Knowing about the immune system always put things in perspective.

So because of this positive effect and just because of the fun of learning and reading about the immune system, it became an ongoing hobby, as I eventually became a science communicator and explaining complex things became my purpose in life. About eight years ago I started Kurzgesagt—In a Nutshell, a YouTube channel dedicated to making information easy to understand and beautiful, while trying to be as true to the science as possible. In early 2021 the Kurzgesagt team has grown to over forty people working on this vision, while the channel has attracted over fourteen million subscribers and reaches about thirty million viewers each month. So if this large platform exists, why go through the horrible process of writing this book? Well, while some of our most successful videos have been about the immune system, it has always bugged me that I could not explore this wonderful topic in the depth it deserves. A ten-minute video is simply not the right medium for that. So this book is a way to turn my decade-long love affair with the immune system into something tangible that will hopefully be a helpful and entertaining way to learn about the stunning and beautiful complexity that makes it possible for you to survive each day.

Unfortunately, the immune system is very complicated, although that is not strong enough a word. The immune system is complicated in the sense that climbing Mount Everest is a nice stroll through nature. It is intuitive like reading the Chinese translation of the tax code of Germany is a fun Sunday afternoon. The immune system is the most complex biological system known to humanity, other than the human brain.

The bigger the immunology textbook you read, the more layers of detail start piling on, the more exceptions to rules appear, the more intricate the system becomes, the more specific it seems to be for every possible eventuality. Every single one of its many parts has multiple jobs and functions and areas of expertise that overlap and influence each other. Even if you make it past these challenges and still want to understand the immune system, you will encounter another problem: The humans who described it.

Scientists have laid the foundation for the amazing modern world we get to enjoy today through hard work and endless curiosity and we owe them a great deal of gratitude. Unfortunately, though, many scientists are really bad at choosing good names and coming up with accessible language for the things they discover. The science of immunology is one of the worst culprits of any scientific discipline in this regard. An already breathtakingly complex field is spiked with words like Major Histocompatibility Complex class I and II, gamma delta T Cells, interferon alpha, beta, gamma, and kappa, and the complement system, with actors named C4b2a3b complex. None of this makes it a pleasure to pick up a textbook and learn about the immune system on your own. But even without this barrier, the complex relationships of the many different actors of the immune system, with countless exceptions and unintuitive rules, are a challenge all by themselves. Immunology is hard even for the people working in public health, even for the people studying immunology, even for the foremost experts in the field.

All of this makes the immune system horrible to explain. If you venture too far into simplification you deprive the learner of the beauty and wonder that lie in the evolutionary genius of the sheer endless complexity that deals with the most crucial problems of living beings. But if you include too much detail, it quickly becomes mind-numbingly hard to keep up with. Listing everything, every part of the immune system, is just too much. It would be like telling someone your whole life story on the first date: Overwhelming and very likely to make them less interested in dating you.

So my aim for this book is to try to carefully dance around all these problems. It will use human language and use complicated words only when necessary. Where appropriate, processes and interactions will be simplified while staying as true to the science as possible. Complexity between chapters will go up and down, so after you are fed a lot of information, there will be more chill parts to relax a bit. And we will summarize what we learned in regular intervals. I want this book to make it possible for everybody to understand their own immune system and have a bit of fun doing so. And since this complexity and beauty are deeply connected to your health and survival, you might actually learn something useful. And of course, the next time you are sick or have to deal with disease, you hopefully can look at your body from a different perspective.

Also, the obligatory disclaimer: I'm not an immunologist, but a science communicator and immune system enthusiast. This book will not make every immunologist happy—what became obvious right from the start of the research is that there are a lot of different ideas and concepts about the details of the immune system and there is a lot of disagreement between the scientists holding these ideas. (Which is how science is supposed to work!) For example, some immunologists consider certain cells useless fossils, while others think they are crucial for your defenses. So as much as possible this book is based on conversations with scientists, the current literature that is used to teach immunology, and peer-reviewed papers.

Still, at some point in the future, parts of this book will need an update. Which is a good thing! The science of immunology is a dynamic field where a lot of amazing things are happening and different theories and ideas are in flux with each other. The immune system is a living topic where great discoveries are still happening. Which is great, because it means we are learning more about ourselves and the world we live in.

OK! Before we jump in and explore what your immune system is doing,

let us define the premise first, so we have solid ground to stand on. What is the immune system, what is the context it works in, and what are the tiny parts that do the actual work? After we have covered these basics we will explore what happens if you hurt yourself and how your immune system rushes in to defend you. Then we'll explore your most vulnerable parts and see how your body scrambles to protect from a serious infection. And lastly, we'll take a look at different immune disorders like allergies and autoimmune disease and discuss how you can boost your immune system. But now let us get to the very beginning of this story.

Part 1

Meet Your Immune System

1 What Is the Immune System?

THE STORY OF THE IMMUNE SYSTEM BEGINS WITH THE STORY OF LIFE IT-self, almost 3.5 billion years ago, in some strange puddle on a hostile and vastly empty planet. We don't know what these first living beings did, or what their deal was, but we know they very soon started to be mean to each other. If you think life is hard because you need to get up early in the morning to get your kids ready for the day, or because your burger is only luke-warm, the first living cells on earth would like a word with you. As they figured out how to transform the chemistry around them into stuff they could use while also acquiring the energy needed to keep going, some of the first cells took a shortcut. Why bother with doing all the work yourself if you could just steal from someone else? Now, there were a number of different ways to do that, like swallowing someone else whole, or ripping holes into them and slurping out their insides. But this could be dangerous, and instead of getting a free meal, you could end up as the meal of your intended victim, especially if they were bigger and stronger than you. So another way to get the prize with less of the risk might be to just get inside them and make yourself comfortable. Eat what they eat and be protected by their warm embrace. Kind of beautiful, if it wasn't so horrible to the host.

As it became a valid strategy to become good at leeching from others, it became an evolutionary necessity to be able to defend yourself against the leeches. And so microorganisms competed and fought each other with the weapons of equals for the next 2.9 billion years. If you had a time machine and went back to marvel at the wonders of this competition, you would be pretty bored, as there was nothing big enough to see other than a few faint films of bacteria on some wet rocks. Earth was a pretty dull place for the first few billion years. Until life made, arguably, the single largest jump in complexity in its history.

We don't know what exactly started the shift from single cells that were mostly on their own to huge collectives working closely together and specializing.*

Around 541 million years ago, multicellular animal life suddenly exploded and became visible. And not only that, it became more and more diverse, extremely quickly. This, of course, created a problem for our newly evolved ancestors. For billions of years the microbes living in their tiny world had competed and fought for space and resources in every ecosystem available. And what are animals really to a bacteria and other critters if not a very nice ecosystem? An ecosystem filled top to bottom with free nutrients. So from the very start intruders and parasites were an existential danger to the existence of multicellular life.

Only multicellular beings that found ways to deal with this threat would survive and get the chance to become even more complex. Unfortunately, since cells and tissues do not really preserve well over hundreds of millions of years, we can't look at immune system fossils. But through the magic of science we can look at the diverse tree of life and the animals that are still around today and study their immune systems. The farther separated two creatures are on the tree of life and still share a trait of the immune system, the older that trait must generally be.

So the great questions are: Where is the immune system different, and what are the common denominators between animals? Today virtually all living beings have some form of internal defense, and as living things become more complex, so do their immune systems. We can learn a lot about the age of the immune system by comparing the defenses in very distantly related animals.

Even on the smallest scale, bacteria possess ways to defend against viruses, as they can't get taken over without a fight. In the animal world, sponges, the most basic and oldest of all animals, which have existed for

---

* Although funnily enough, it may actually have been a side effect of single-celled organisms being mean to each other. As at one point one cell swallowed another but did not devour it. Instead these two cells started arguably the most successful partnership on planet Earth that is still going strong today. The "inside cell" (that we know as "mitochondria" today) specialized in making energy available for the host, while the "outside cell" offered protection and delivered free food. This deal worked very well and enabled the new super cell to grow in complexity and become more and more sophisticated.

more than half a billion years, possess something that probably was the first primitive immune response in animals. It is called *humoral immunity*. "Humor," in this context, is an ancient Greek term that means "bodily fluids." So humoral immunity is very tiny stuff, made out of proteins, that floats through the bodily fluids outside of the cells of an animal. These proteins hurt and kill microorganisms that have no business being there. This type of defense was so successful and useful that virtually all animals around today have it, including you, so evolution did not phase this system out, but rather, made it crucial to any immune defense. In principle, it hasn't changed in half a billion years.

But this was only the start. Being a multicellular animal has the perk of being able to employ many different specialized cells. So it probably did not take animals too long, in evolutionary terms, to get cells that did just that: Specialize in defense. This new *cell-mediated immunity* was a success story right from the start. Even in worms and insects we find specialized soldier immune cells that move freely through the tiny critter bodies and can fight intruders head-on. The further up we climb the evolutionary tree, the more sophisticated the immune system becomes. But already, on the earliest branch of the vertebrate part of the tree of life, we see major innovations: The first dedicated immune organs and cell training centers, together with the emergence of one of the most powerful principles of immunity—the ability to recognize specific enemies and quickly produce a lot of dedicated weapons against them, and then to remember them in the future!

Even the most primitive vertebrates, jawless fish, who look ridiculous, have these mechanisms available to them. Over hundreds of millions of years, these defense systems got more and more sophisticated and refined. But in a nutshell, these are the basic principles, and they work well enough that they were probably around in some forms around half a billion years ago. So while the defenses you have at your disposal today are pretty great and developed, the underlying mechanisms are extremely widespread and their origins reach back hundreds of millions of years. Evolution did not have to reinvent the immune system over and over again—it found a great system and then refined it.

Which finally brings us to humanity. And to you. You get to enjoy the fruits of hundreds of millions of years of immune system refinement. You are the height of immune system development. Although, your immune system is not really inside of you. *It is you.* It is an expression of your biology protecting itself and making your life possible. So when we are talking about your immune system, we are talking about *you.*

But your immune system is also not a singular thing. It is a complex and interconnected collection of hundreds of bases and recruitment centers all over your body. They are connected by a superhighway, a network of vessels, similarly vast and omnipresent as your cardiovascular system. Even more, there is a dedicated immune organ in your chest, as big as a chicken wing, that gets less efficient as you age.

On top of organs and infrastructure, dozens of billions of immune cells patrol either these superhighways or your bloodstream and are ready to engage your enemies when called. Billions more sit guard in the tissue of your body that borders your outsides waiting for invaders to cross them. On top of these active defenses you have other defense systems made up of quintillions of protein weapons that you can think of as self-assembling, free-floating land mines. Your immune system also has dedicated universities where cells learn who to fight and how. It possesses something like the largest biological library in the universe, able to identify and remember every possible invader that you may ever encounter in your life.

At its very core, the immune system is a tool to distinguish the **other** from the **self.** It does not matter if the other means to harm you or not. If the other is not on a very exclusive guest list that grants free passage, it has to be attacked and destroyed because the other might harm you. In the world of the immune system, any "other" is not a risk worth taking. Without this commitment you would die within days. And as we will learn later, sadly, when your immune system under- or overcommits, death or suffering are the consequences.

While identifying what is self and what is other is the *core,* it is not technically the *goal* of your immune system. The goal above all things is maintaining and establishing *homeostasis*: the equilibrium between all the elements and cells in the body. Something that can't be overemphasized enough about the immune system is how much it tries to be balanced and how much care it puts into calming itself down and not overreacting. Peace, if you so want. A stable order that makes being alive pleasant and easy. The thing that we call health. The basis for a good and free life where we can do what we desire, not held back by pain and disease.

How crucial health is becomes the most apparent when it is missing. Health is really an abstract concept because it describes the absence of something. The absence of suffering and pain, the absence of limitations. If you are healthy, you feel normal, you feel right. Once you witness your health go away, even for a brief time, it is hard to forget how fragile you are and how much you are living on borrowed time. Disease is an unavoidable fact of life. If you have been lucky, you have not had to face it up to this point. If you or one of your loved ones has had to deal with it already, you know that nothing is more elemental for a pleasant life than being healthy. To the immune system, this means homeostasis. While the battle to stay healthy is ultimately futile and will be lost in the end, we still fight it to carve out more years, months, days, and hours. Because, overall, it is pretty good to be a human and it is worth it to have this experience a little bit longer.

But health is a hard thing to maintain because every day of your life you are in contact with hundreds of millions of bacteria and viruses that would love to make your body their home, as we saw in those single-celled organisms billions of years ago. For a microorganism you are an ecosystem waiting to be conquered. An endless continent full of resources, breeding grounds, and opportunities to thrive, a really nice home. Arguably at some point they will succeed, as when you die, the decomposition of your body will be immensely sped up by an army of unhinged microbes no longer kept in check by your defenses.

And not only do you need to worry about the plethora of life trying to get inside, but also about your own misguided self that can cancel the social contract of the body: Cancer. Making sure that doesn't happen is one of the most important jobs of your immune system. In fact, while you were reading the last few pages, somewhere inside you a young cancer cell was quietly eliminated by your immune cells.

But the part meant to protect you also can go wrong and be corrupted. When it is tricked, your immune system can help diseases spread or protect cancer cells from detection. Or if the system is out of tune or flawed, it can get confused and decide that the body itself is the enemy. It can decide that *self* is *other* and literally start attacking the cells it exists to protect, resulting

7

in any number of autoimmune diseases that need constant calming by medication, sometimes with harsh side effects.

Or take allergies, which are a very intense reaction of your immune system against things it should not be concerned about. An allergic shock shows strikingly how truly powerful your defense system is and how horribly it can go wrong: it may take a disease days to kill you—your immune system can do so in minutes.

Oh, and even if your immune system works as intended, it can be as much of a burden as it is helpful: Many of the unpleasant symptoms you feel when you are sick are the consequences of your immune system doing its job when activated—in some diseases the most crushing damage or even death are caused by an unhinged response to an intrusion. For example, many deaths from COVID-19 come from the immune system doing its job with too much enthusiasm.

The collateral damage your defense networks dish out against you can build up over time and today it is thought that many deadly diseases start with your immune system working as intended. So as important as it is for your health to have an immune system that is fast and brutal, it is just as important to keep it in check and prevent it from becoming unhinged and destructive. Just like in the human world, if you need to go to war, you at least want your wars to be over quickly and end with a clean victory. You don't want decades of occupation or conflict that eat up resources and leave destroyed infrastructure.

So in the hands of your immune system lies an enormous responsibility to keep you well for as long as possible. Even if the battle will certainly be lost in the end, it matters to you today, right now, that it is fought well and with the necessary responsibility.

To summarize, distinguishing between self and other is core, homeostasis is the goal, and there are seemingly infinite ways for it to all go wrong.

What makes the immune system so fascinating is that all of this complex work has to be done by parts that are mindless and, individually, pretty dumb. And yet they are able to coordinate and react to dynamic and quickly developing situations. Imagine the Second World War happening, but ten times as large and without generals. Only mindless immune soldiers on the ground trying to figure out if they need tanks or fighter jets and where they need to go. And it all happens within days. That's what it is like for you to battle even a common cold.

So now let us discover your immune system, so the next time you step into the shower, annoyed about the cold symptoms you are feeling, you can at least pause for a moment to appreciate what is going on inside you before you go back to being annoyed.

2 What Is There to Defend?

BEFORE WE CAN REALLY LEARN ABOUT YOUR INTRICATE DEFENSE SYSTEM, we should take a look at what needs to be defended: Your body. In a sense this seems pretty straightforward—it is the entire area under and including your skin. Simple enough, right? But just like looking at a planet in space, you will never see anything remotely approaching the full picture from orbit.

So before we do anything else, first we need to go on a journey together, into a strange and foreign world, stranger than the deep sea or an alien planet. A world where no living being therein even knows that it exists, where monsters are a daily reality but nobody cares. A world billions of years old, that exists within yourself, within everybody and everything, all around us, omnipresent but invisible. This is the world of the tiny, where the border between dead and alive becomes fuzzy. Where biochemistry becomes life for reasons we still do not understand. Let us zoom in to you and take a look—into your organs, through the tissue, to our most fundamental building blocks, your cells.

Cells are extremely tiny, living things, among the smallest units of life on earth. For a single cell, your body is a planet drifting through a hostile universe. To understand the enormous dimensions of your body we need to look at it from a cell's perspective. At the scale of a cell, your body is an utterly gigantic structure of pipes as wide as mountains, filled with oceans of fluids, rapid torrents permeating intricate cave systems stretching as far as whole countries. With the exception of the crystallized and hard parts of your bones, all of the environment, all of the world, really, is, to a cell, alive. A cell can politely ask a wall to let it pass and then squeeze through a tiny gap that closes behind it. It can swim through channels and hike up mountains of meat to get anywhere it needs to go.

If you were the size of one of your cells, the body of a human would be in the area of fifteen to twenty Mount Everests stacked on top of each other. It would be a mountain of flesh at least sixty miles (a hundred kilometers) tall, reaching into space. If you are near a window take a second and look into the sky. Try to imagine this for a moment, a giant so large that passenger planes would crash against its lower legs, its head so far above you that you would not be able to see it.

The cells of your immune system are tasked to defend all of *that*. Especially the weak points where intruders can enter it, which is mostly the borders, the *outsides* of the body. When you think about your *outsides*, the first thing that comes to mind is, of course, your skin. The total surface area of your skin is about two square yards (meters) (about half the size of a pool table) and luckily is not that hard to defend, since most of it is made out of a hard and thick barrier covered with its own defense system. It feels soft, but it is pretty hard to breach if it is intact.

Your *real* weak points to infections are your mucous membranes—the surface that lines your windpipe and lungs, eyelids, mouth, and nose, your stomach and intestines, your reproductive tracts and bladder. It is hard to give their total surface area since numbers vary so much from person to person, but on average there are about 200 square yards (meters) of mucous membranes in a healthy adult (about the same as a tennis court), most of them being your lungs and your digestive tract.

You may mistakenly think of your mucous membranes as your insides. But this is not true—your mucous membranes are outsides. If we took an honest look at what you are, you are, in a sense, nothing more than a complex tube. Granted, a tube that can close both ends. Also a lot wetter, slimier, and grosser.

Your reproductive organs, nostrils, and ears are extra holes—entrances to large tunnels and additional cave systems that reach through you. All of these places are your direct borders, contact points with the outside world. Your body is just wrapped around them. These outsides, within your insides, represent surfaces where millions of intruders are trying to enter you every day. A lot of ground to defend when you are the size of a cell. For your cells, the surface area of the mucous membranes is as big as Central Europe or the central United States are to you. Building border walls would not do

11

Your body is a tube.

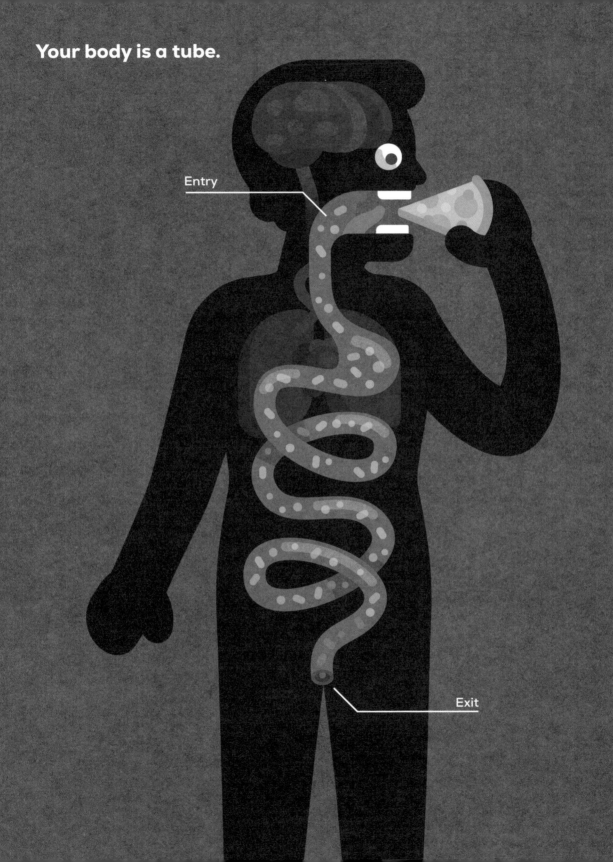

Entry

Exit

the trick for them, since they don't need to defend just the borders, but the *entire surface*! It's not like intruders are trying to enter just at the edges. They could sort of drop in with parachutes. So your cells need to defend the entire continent. All of it.

Still, it's much easier to catch an enemy at one of these points than somewhere else—for example, if we took all the blood vessels and capillaries from your body and laid them out in a straight line, they would be a baffling 10,000 miles (16,000 kilometers) long—three times the circumference of Earth—with about 1,400 square yards (1,200 square meters) of surface area. So, better catch enemies at the borders that are significantly smaller and therefore easier to defend. But easier does not mean easy.

Let's do a fun experiment and imagine that we wanted to build a human body to scale but from actual people, like you. Living breathing humans, just to see what sort of crazy dimensions we encounter here.

So first, we need a lot of people for that. The average human body is made from around forty trillion cells. *Trillion!* Forty trillion is 40,000,000,000,000. A truly impressive number. If we want your cells to be represented by individual people, then we need more than one hundred times as many people as have lived in the 250,000-year-long history of humanity. Let's try to visualize this a bit. Right now around 7.8 billion people are alive. If we put them shoulder to shoulder, they surprisingly would only cover an area of around 700 square miles (1,800 square kilometers). Which is a little bit more than the surface area of London. To get forty trillion people we need to multiply this by 120.*

---

* And this is only half of the story because your body hosts bacteria that you need to survive. How many? One bacterium for each of the forty trillion cells in your body (which happens to be a pretty good ballpark in terms of size; if you were the size of an average body cell, a bacterium would be roughly the size of a bunny). Let us imagine them as baby bunnies to make the thought less horrifying. Most of these cute bunnies live in your gut. In this ginormous cave, 36 trillion bunnies live their lives, constantly dying and reproducing, breaking down chunks of food the size of skyscrapers so it can be distributed to all the people making up the continent of flesh. The other four trillion bunnies are crawling on your skin, are inside your lungs, hopping over your teeth and your tongue, they swim in the fluid of your eyes, crawl in and out of your ears. We'll talk more about them later but for now just imagine yourself as being covered by cute bunnies who are your friends and have only your best interests in mind.

All humans alive ~7.8 Billion

•

Cells in your body ~40 Trillion

• = 10 Billion

OK. So we now have forty trillion people, standing shoulder to shoulder. This ocean of people would cover the whole of the United Kingdom, every last corner, lake, and mountain. To make a body to scale, made of people representing cells, we need to stack them until trillions of people are standing on top of each other, holding hands and linking arms, forming living structures. A giant made out of flesh rises 60 miles (100 kilometers) into the sky, reaching the edge of space. The giant is made up of caverns as wide as small countries, bones as dense and wide as mountains, filled with intricate caves and tunnels. Its arteries are filled with oceans of fluid and people carrying food and oxygen tanks to every last corner. If you were a red blood cell, or in this case "red blood person," you would travel the distance from Paris to Rome and back once a minute in a stream pumped by a heart as large as a city. Things could be great. Everybody would work together to keep the mountain of flesh, and in consequence, themselves, alive.

But the enormous richness of resources and food and the abundance of moist, warm space is just too attractive. The giant is not only the size of a continent for its inhabitants but also for unwelcome visitors. Literally billions of parasites are trying to get inside the flesh giant. Some are as large as elephants or blue whales and they want to lay giant eggs so their young can feast on the poor people that make up the tissues. Others are the size of raccoons or rats who want to steal the food and make the giant their permanent home to raise generations of their offspring. They may not intend harm to the people making up the body, but they will do so by defecating everywhere, making life miserable. The most disgusting vermin our flesh giant has to deal with daily are billions of spiders who want to enter the cell people's mouths or ears to breed in the stomachs of their victims. For a giant made up of trillions of people, losing a few here and there is not really dangerous. But if the vermin were allowed to procreate freely, it could be its end. Isn't the idea horrifying?

This is what your cells deal with every single day and night, from your birth until the day you die. Staying alive is not a thing you should take for granted. But don't let this idea of being attacked distress you too much. You are not just a mountain of flesh waiting to be conquered. Thankfully you have a great ally in this fight for survival that, as we now know, we simply don't cherish and celebrate as much as it deserves: your immune system.

It makes you a fortress. And even more, a fortress filled with billions of the most effective and fierce soldiers in the universe. They have countless weapons at their disposal and they use them without mercy. Your immune system army has already killed billions of enemies and parasites in your life, and it is ready to kill billions or trillions more.

3 What Are Your Cells?

We've talked an awful lot about cells so far and will do so even more in the rest of this book. In order to understand your body, your immune system, and the diseases it fights, from cancer to the flu, you need a base understanding of its building blocks. It helps that cells are maybe the most fascinating part of biology. After this chapter we will zoom out again and meet your immune system in earnest.

So what exactly is a cell and how does it work?

As we said, cells are the smallest units of life: things that we can clearly identify as something that is alive. The definition of life is a big, complicated, brain-melting affair in itself. We know it when we see it but it is very hard to define. In general there are a few properties that we assign to it: Something alive separates itself from the universe around it. It has a metabolism, meaning it takes up nutrients from the outside and gets rid of internal garbage. It responds to stimuli. It grows and it can make more of itself. Cells do all these things. And you are made almost entirely from them. Your muscles, organs, skin, and hair are made up of cells. Your blood is filled with them. Being as small as they are, they are not conscious, they do not have free will or feelings or goals or make active decisions. In a nutshell, cells are biological robots, driven entirely by myriads of biochemical reactions guided by the even smaller parts they are made up from.

Your cells have "organs" that are called organelles, like the nucleus, the information center of your cell—a pretty large structure with its own protective border wall that houses your DNA, your genetic code. There are mitochondria, generators that transform food and oxygen into chemical energy that keeps your cells running. There is a specialized transport network, a packaging center, parts for digestion and recycling, construction centers. When we learn about cells, they are often illustrated as sort of empty bags filled with these organelles. But this image gives the wrong impression of how much they are buzzing with complex activity. Look around the room you are sitting in right now.*

Now imagine the room to be filled top to bottom with stuff. Millions of grains of sand, millions of grains of rice, and a few thousand apples and peaches and a dozen large watermelons. This is sort of what the inside of a cell looks like. What does this mean in reality?

A single human cell is filled up with dozens of millions of individual molecules. Half of them are water molecules, in our metaphor represented by the grains of sand that give the insides of your cells a consistency that is kind of like soft jelly and enable other things to move around easily. Because on this scale water is no longer a thin fluid but viscous and honey-like.†

The other half of your cells' insides consists mostly of millions of proteins. Between 1,000 and 10,000 different kinds—depending on the function of the cell and what it needs to get done. In our room example, they would be the rice and most of the fruits. The watermelons are the organelles that we always see in pictures of cells. So your cells are mostly made from and filled up with proteins.

We need to briefly talk about *proteins* because they are super important to understanding the immune system and your cells and the microworld they live in. They are so important that it is OK to call your cells protein robots. You may have heard of proteins mostly in the context of food—maybe you even are on a protein-rich diet, especially if you are working out a lot and are trying to build up muscle. Which makes sense because the solid, nonfat parts of your body are mostly made out of protein (even your bones are made from a mix of proteins and calcium). But proteins are not just good for muscles: Proteins are the most fundamental organic building blocks and tools of all living things on this planet. They are so useful and manifold that a cell

---

* If you are reading this outside, well, that is unfortunate for the metaphor, isn't it? So please pretend you are somewhere inside.

† You may ask yourself why that is. Well, we could spend a lot of time talking about this and it is actually quite fascinating but also opening a whole new can of worms. So let us just say that it matters how big you are. While for you on the human scale water is this uniform substance, if you were the size of a protein, a single water molecule is pretty big, like an actual thing that bumps into you. So you would find it much harder to swim through water too.

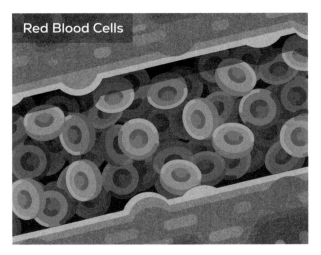

Red Blood Cells

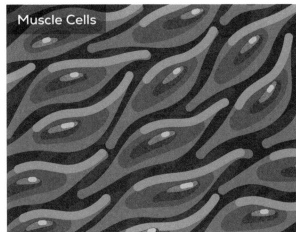

Muscle Cells

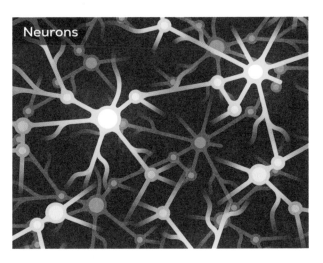

Neurons

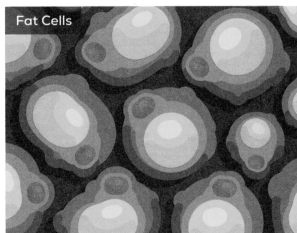

Fat Cells

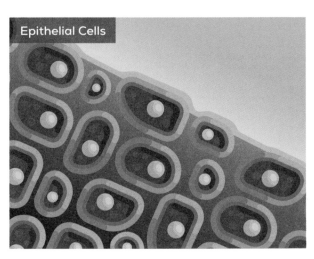

Epithelial Cells

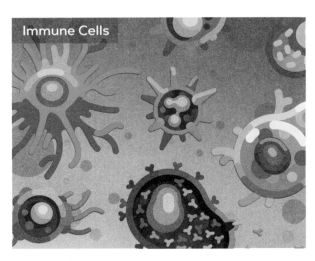

Immune Cells

can use them for basically everything, from sending signals to constructing simple walls and structures to complex micromachines.

Proteins are made from chains of amino acids, which are tiny organic building blocks that come in twenty different varieties. All you need to do is to string them together into a chain, in whatever order you like, and voila, you have a protein. This principle enables life to construct a stunning variety of different things. For example, if you wanted to make a simple protein from a chain of ten amino acids, and you have twenty different amino acid types that you can choose from, this gives you a breathtaking 10,240,000,000,000 different possible proteins.

Imagine having a casino slot machine, with twenty different symbols and ten different slots. It is hard enough to get the same symbol on a machine with three slots—imagine how many possible combinations on your protein slot machine. A typical protein is usually made from between 50 and 2,000 amino acids (which would be the equivalent of a slot machine with 50 to 2,000 slots) and the longest ones we know of are made from up to 30,000. This gives us billions of billions of potentially useful proteins our cells can make.

Of course, most of these possible proteins will be useless. According to some estimates, only one in a million to a billion possible amino acid combinations will yield a useful protein. But since there are so many possible proteins, one in a billion is still plenty! How do your cells know in which order to put amino acids to make the proteins they need?

Well, this is the job of the code of life: Your *DNA*, a long sequence of instructions that are necessary for a living thing to be a living thing. What this means in this context is that around 1% of the DNA is made up of sequences that are building manuals for proteins, which are called *Genes*. The rest of your DNA is regulating which proteins are built when and how and how many of them at which time. So proteins are so crucial to living beings that the code of life is basically an instruction manual for building them. But how does this work? Well, very briefly, and only because this will become important later on when we talk about viruses: In a nutshell, the instructions on the DNA are converted into proteins in a two-step process: Special proteins read the information on the DNA string and convert it into a special messenger molecule called mRNA—basically the language that our DNA uses to communicate orders.

The mRNA molecule is then transported from the nucleus of the cell to another organelle, the protein production machinery called the ribosome. Here the mRNA molecule is read and translated into amino acids, that are then put together in the order inscribed into it. And voila, the cell has made a protein from your DNA. So your DNA is basically a bunch of code, with sections called genes, that are a protein-building and regulation manual for your cellular machinery. And this does actually translate into all the features that you as an individual call your own: Your height, your eye color, how susceptible to certain diseases you are, or if you have curly hair. Your DNA doesn't tell your body "make curly hair!"—it tells your cells "make these proteins." In a really simplified sense, all of your personal traits manifest this way.

You have a lot of this genetic code—if you were to untangle the DNA of a *single* one of your cells, it would be about two meters long. That's right, the DNA inside every one of your cells is in all likelihood longer than you are tall. If we took all of your DNA from your body and combined it into one long string, it would reach from Earth to Pluto and back. And all of that code just to make long chains of amino acids!*

As these amino acid chains are made, they transform and change from a long 2D string into a 3D structure. This means that they are folding in on themselves in really complicated ways that we haven't fully deciphered yet. Depending on the types of amino acids and the sequence they are put together in, the chain folds together into specific shapes.

In the world of proteins shape determines what they can and can't do. Shape is everything. In a way, you can imagine proteins like really complex

---

* Some of you will do the math now and get even crazier numbers. Forty trillion cells times two meters is roughly 80,000,000,000,000 meters, which is actually five times the distance to Pluto and back to Earth. But there is a minor catch we didn't mention in the intro about the body: The vast majority of your cells actually do not have DNA. Red blood cells in particular make up around 80% of your cells by pure numbers, and they don't have a nucleus, because they are filled head to toe with iron molecules that transport oxygen. So you will have to settle for going to Pluto and coming back just once.

Proteins

Proteins are the most common
construction material for your cells.
But they also transmit messages or
convey information. Cells can build
basically everything from proteins.

Pepsin

Actin

Antibody

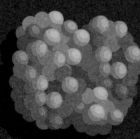

Glutamine Synthetase

Hemoglobin

10 Nanometers

three-dimensional puzzle pieces. Depending on their shape, proteins are the ultimate tool and construction material. A cell can use them to build basically everything. But the magic of proteins goes beyond being merely construction materials. Proteins are used as messengers that convey information: They can receive or send signals that change their shape and trigger intensely complicated chain reactions. For your cells, proteins are everything. Think back to the room filled with rice and peaches and apples. All these proteins are actually not like spheres but more like an unfathomably complex mix of gears and wheels and switches and domino pieces and tracks.

As long as your cell is alive it is always moving and shifting. Wheels spin and tip over dominos, which push switches and pull levers and ferry marbles around on tracks that then spin more wheels, and so on. If you want to get metaphysical, the spirit of the cell robot is both the proteins and the biochemistry guiding them.

Some of the most common proteins are extremely plentiful inside your cells, with up to half a million individual copies. Others are specialized and exist fewer than ten times in total. But they don't just float around doing their own thing. All these tiny little protein puzzle pieces and structures inside your cells interact in lots of really cool and complex ways. How do they do that? By wiggling around really fast. Proteins are so small, weigh so little, and exist on such a fundamentally different scale that they behave very strangely compared to things on the human giant level. Gravity is not a relevant force for things at this scale. And so, at room temperature, an average protein can move about sixteen feet (five meters) per second, in theory. Maybe that does not sound fast, until you remember that the average protein is about one million times smaller than the tip of your finger. If you could run as fast as a protein in your world, you would be as fast as a jet plane and die horrifically by crashing into something.

In practice, proteins can't actually move that fast inside cells, because there are so many other molecules in the way. So they constantly collide and bump into the water molecules and other proteins in all directions. Everybody is pushing around and is getting pushed. This process is called *Brownian motion* and it describes the random movement of molecules in a gas or fluid. Which is the reason water is so important for your cells—because it enables other molecules to move around easily. Despite, or maybe even because of the chaos of random movements in combination with the speed of your protein puzzle pieces, things get done in cells.*

Let's try to simplify a little. To imagine the basic principle cells use to bring things together, a good metaphor is a sandwich. If you were inside a cell and you wanted to make a jelly sandwich, the best approach would be to throw the toast and jelly into the air and wait a few seconds. Because of how fast everything is smashing together, they will come together all by themselves, unifying into a sandwich that you can just pick from the air.†

In the microworld, the different shapes of a molecule determine which molecules attract and repel each other. And so the shape of your cells' proteins determines which proteins attract or repel each other and how they interact (while the number of different types of proteins determines how often these interactions happen). This creates the interactions that make up the biochemistry of all cells on earth. These interactions are fundamentally important for biology and are called *biological pathways.* Pathways are fancy words to describe a series of interactions between individual things that lead to a change in a cell. This can mean the assembly of new special proteins or other molecules, which can turn genes on and off, which changes what the cell can and can't do. Or it can spur a cell into action and cause the cell to do things we would call *behavior,* like reacting to a danger by moving away.

OK. That was a lot of information in the last few pages. And we aren't quite out of the cell yet, but almost! Let's summarize quickly what we learned:

Cells are filled up by proteins. Proteins are three-dimensional puzzle pieces. Their specific shapes enable them to fit together or interact with other proteins in specific ways. Sequences of these interactions, called pathways, cause cells to do

---

* This is not to say that our complex human cells are completely reliant on randomness. Cells have a lot of complex and wonderful mechanisms to get things exactly where they need them to be, and that we'll ignore here. If you do care: There are transport proteins that move along the scaffolding of cells. The best thing about them is that they look like gigantic, ridiculous feet that jump forward by magic and if you have a moment to get distracted, you should look at videos of them on YouTube.

† In reality it is more like throwing thousands of toasts and thousands of pots of jelly into the air. Your cells have no use for a single jelly sandwich but need large amounts of everything to make things work out.

things. This is what we mean when we say that cells are protein robots guided by biochemistry. The complex interactions between dumb and dead proteins create a less dumb and less dead cell, and the complex interactions between slightly dumb cells create the pretty smart immune system.

As is the case with most of this stuff, we stumbled into a big topic here and there are countless rabbit holes to fall into. In this case, we stumbled into how and why many mindless things can create something that is smarter than the sum of its parts. This is typically not discussed when the immune system is explained, but it might be worth spending a minute on it before we move on, because it adds another layer of wonder to the immune system and your cells in general that we never really think about when we have to sit out a flu or watch a wound heal.

But because this all gets abstract quickly, we need another analogy, so let us talk about ants for a hot second. Ants share a few properties with cells, most importantly: they are really dumb. This is not to be mean to ants. If you take a single ant and isolate it, it will just stumble around and be really useless, unable to do anything of value. But if you put a lot of ants together, they can exchange information and interact with each other and in unison do amazing things. Many ants construct complex structures with specialized areas like brood chambers, dedicated garbage places, or complex ventilation systems that control airflow. Ants automatically organize themselves into different classes and jobs, from foraging to defending or nursing. Not just randomly, but in proportions that are the most useful for the survival of the collective. If one of these different classes is decimated, maybe because of a hungry anteater passing by, some of the remaining ants will switch their job to restore the correct job ratio again. And they do all of these things despite being really dumb individually. But together, they become something greater and are able to do legitimately astonishing things that they could not do alone. This phenomenon occurs all over nature, and is called *emergence*. It is the observation that entities have properties and abilities that their parts do not have. So an ant colony as an entity can do complex things, while the individual ant can't.

This is sort of how everything in your body works. Your cells are nothing but bags of proteins guided by chemistry. But together these proteins form a living being that can do a lot of really sophisticated things. Still, our cells remain mindless robots that are, individually, even dumber than ants. But many of them acting together can do things the individuals can't do. Like forming specialized tissue and organ systems, from muscles that make your heart beat to brain cells that make you think and read this sentence. And many stupid parts and cells together form your immune system—through complex interactions that turn out to create something really smart.

OK, we have to move on. But from this diversion you hopefully gathered the following things: Cells are the wonderfully complex machines of life. They are mostly made from and filled with puzzle pieces made from a stunningly diverse number of proteins, and entirely run by biochemistry. Somehow all of this together creates a living thing that can sense and interact with its environment. Cells do their jobs without emotion or purpose. But they do them very well and for that they deserve our gratitude and a little bit of attention. In the following chapters we will anthropomorphize our tiny robot cells from time to time.

We will speak about what they want and are trying to achieve, their thoughts, hopes, and dreams. Which gives them a little bit of character and makes it easier to explain certain things, even if it is not true. As amazing as your cells are, please remember: Cells don't want anything. Cells don't feel anything. They are never sad or happy. They just are, right here, right now. They are as conscious as a stone or a chair or a neutron star. Cell robots follow their code that has been evolving and changing for billions of years and has turned out pretty great if you are able to sit down comfortably and read this book right now. Still, seeing them as little buddies might lead to us treating them with more respect and understanding and will make this book much more fun to read, which seems like a good enough excuse to do so.

Now, you may ask yourself: If we have this huge continent of flesh that is populated by billions of robots that are collectively smart while being, on an individual scale, internally complex yet pretty stupid—how on earth do they defend your body?

Well . . .

4 The Empires and Kingdoms of the Immune System

IMAGINE YOU WERE THE GRAND ARCHITECT OF THE IMMUNE SYSTEM. YOUR job is to organize the defenses against millions of intruders that want to take it over. You get to build whatever defenses you like, although the accountants remind you that the body is on a tight energy budget, has no resources to spare, and they kindly ask you to not be wasteful. How would you approach this monumental task? What kind of forces would you put at the front and which ones would you hold in reserve? How would you make sure that you could react strongly to a sudden invasion but also prevent your army from exhausting itself too quickly? How would you deal with the massive scope of the body and the millions of different enemies you would have to account for? Luckily, your immune system has found many beautiful and elegant solutions for these problems.

As we alluded to in the last chapter, the immune system is not a singular thing but many different things. Hundreds of tiny organs and a few bigger ones, a network of vessels and tissues, billions of cells with dozens of specializations and quintillions of free-floating proteins.*

All these parts form different and overlapping layers and systems, so it's helpful to imagine them as empires and kingdoms that, in unison, defend the continent that is your body. We can organize them into two very different realms that together represent the most powerful and ingenious principles

---

* So you have probably heard that you have white blood cells and they are your immune cells or something like that. Well, while this name has its use in the right context it just generally means "the cells of the immune system" and I don't think immunology has done itself a favor with this term. "White blood cells" describes so many different cells that do so many different things that it is sort of useless if you want to understand what is really going on here. So you can forget "white blood cells" again because we are not going to use it.

Most Important Players of the Immune System

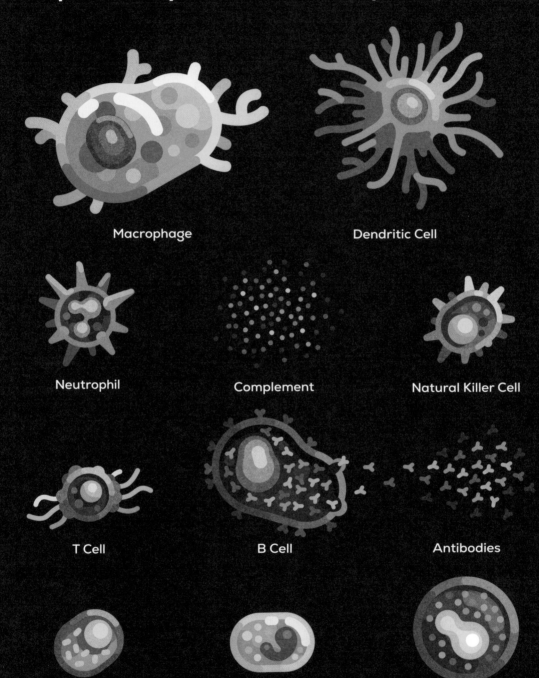

Macrophage

Dendritic Cell

Neutrophil

Complement

Natural Killer Cell

T Cell

B Cell

Antibodies

Basophil

Eosinophil

Mast Cell

that nature found to defend your continent of flesh: The Realm of your **Innate Immune System** and the Realm of your **Adaptive Immune System**.

The Realm of the Innate Immune System contains all the defenses you are born with and that can be employed mere seconds after an invasion occurs. These are the basic defenses that go back to the very first multicellular animals on earth and they are absolutely crucial for your survival. One of its most central features is that it is the sort of smart part of your immune system. It has the power to tell *self* from *other*. And once it detects *other* it immediately springs into action. However, its weapons are not tailored to identify any specific enemy, but instead they try to be effective across a wide range of common enemies. It doesn't have specific weapons against specific types of *E. coli* bacteria, for example, but against bacteria in general. It's designed to be as widely effective as possible. Think of it like your basic starter kit: it has all the fundamentals, not the specialized items you'd get with an advanced kit. But without the fundamentals, the specialized items are all but useless.

Without your Innate Immune System, you would be overwhelmed and killed by microorganisms within days or weeks. It does the heavy lifting and most of the actual fighting. The vast majority of your hundreds of billions of soldier and guard cells are part of your innate immune system. These are pretty rough fellows that prefer bashing in heads over talking and thinking. Most microorganisms that successfully invade you are killed by your innate immune system without you even noticing. Since the Innate Immune System is the first line of defense it is not just responsible for throwing soldiers at danger, it also has to make crucial decisions: How dangerous an invasion? What kind of enemy is attacking? And are more heavy weapons necessary?

These decisions are vital because they influence what sort of weapons your immune system as a whole will deploy. A bacterial invasion needs a very different response than a viral invasion. So while a fight is going on, the Innate Immune System gathers intel and data and then it makes the decisions that in many cases will decide your fate. If your innate immune system thinks an attack is serious enough, it has the power to activate and call the second line of defense to mobilize and join the fight.

The Realm of the Adaptive Immune System contains specialized super cells that coordinate and support your first line of defense. It contains factories that produce heavy protein weapons and special cells that hunt and kill infected body cells in the case of viral infections. Its defining feature is that it is *specific*. Unbelievably specific, in fact. Your Adaptive Immune System "knows" every possible intruder. Its name, what it had for breakfast, its favorite color, its most intimate hopes and dreams. The Adaptive Immune System has a specific answer for every single possible microorganism that exists on this planet right now—and for every single one that can evolve in the future. Think about how creepy that actually is. If you were a bacteria, for example, all you would want is to get into a human and find a place to make babies but suddenly there are agents that know your name, your face, your personal history, and all of your most intimate secrets and they are armed to the teeth.

This breathtakingly specific defense and how it works will be a focus of future chapters, but for now, just remember that your Adaptive Immune System possesses the largest library in the known universe, with an entry for every current and future possible enemy. But not only that, it also is able to remember everything about an enemy that showed up only once. It is the reason most diseases are only able to manifest themselves once in your life. But this knowledge and complexity come with downsides.

In contrast to the Innate Immune System, your Adaptive Immune System is not ready yet when you are born. It needs to be trained and refined over many years. It starts as a blank slate and then gets progressively more powerful, only to get weaker again as you age. A weak Adaptive Immune System is one of the main reasons young and old humans are often much more likely to die from diseases than people in the middle of their lives. Mothers actually lend their newborn babies a bit of their adaptive immunity in their mother's milk to help them survive and give them some protection!

While it is easy to think of the Adaptive Immune System as your more sophisticated defense, one of the most important things it actually does is to make your innate defenses stronger by motivating your innate soldier cells to fight harder and more efficiently (but more on that later).

For now, let's summarize: Your Immune System consists of two major realms: Innate and Adaptive Immunity. Your Innate Immune System is ready to fight after birth, and can identify if an enemy is not *self*, but *other*. It does the down-and-dirty hand-to-hand combat, but it also determines what broad category your enemies fall in and how dangerous they are. And finally it has the power to activate your second line of defense: Your Adaptive Immune System, which needs a few years before it is ready to deploy efficiently. It is *specific* and can draw from an incredibly large library to fight every possible individual enemy that nature can throw at it, with powerful superweapons. But while it is powerful, one of its most important jobs is to make the Innate Immune System even stronger.

Both of these realms are interconnected in a deep and stunningly complex way. And it is in the interactions between these two systems wherein lies some of the magic and beauty of your immune system.

To explore the different realms with the attention they deserve, the rest of this book is organized into three major parts. In part 2 we will experience an invasion that will occur through your skin and by bacteria and in part 3 we will witness a sneaky surprise attack on your mucosa by viruses. In part 4 we will then see how everything comes together and discuss specific disorders and diseases, from autoimmune to cancer.

So now let us see what happens if your borders are breached.

Part 2

Catastrophic Damage

5 Meet Your Enemies

To understand your defenses, it is critically important to understand who is attacking you. As we said before, to most living things you are not a person but a landscape covered by forests, swamps, and oceans filled with rich resources and plenty of space to start a family and settle down. You are a planet, a home.

Most of the microorganisms that accidentally enter your body are dealt with rather quickly as they are simply not prepared for the body's harsh defense measures. So the majority of the living things surrounding you are just mildly annoying to your immune system.

Your true enemies are an elite group that has found ways to overcome your defenses more effectively. Some even specialize in hunting humans, or are using you as a crucial part of their life cycle—enemies like the measles virus, for example, that have decided to be super annoying to us. Or *Mycobacterium tuberculosis,* which may have co-evolved with us as far back as 70,000 years ago and still kills about two million people each year. Others, like the novel coronavirus that causes the disease COVID-19, stumble onto us by accident and can't believe their luck.

In today's modern world when we are thinking of things that make us sick, we are mostly talking about bacteria and viruses. Although, in developing countries, protozoa, single-celled "animals" that cause diseases like malaria, which kills up to half a million people each year, are still a serious problem.

Any sort of invader that is able to give your immune system a run for its money is called a **Pathogen**—which appropriately means "the maker of suffering." So every microorganism that causes disease is a pathogen, no matter what species, no matter how big or small. And almost everything can become a pathogen under the right circumstances. For example, a regular old bacteria living on your skin might not bother you at all, but can become a pathogen if you are going through chemotherapy and are immunocompromised, making it easy to invade you. So whenever you read "pathogen," just remember it means "a thing that makes you sick."

Your immune system is "aware" that there are very different kinds of pathogens, that all require very different responses to get rid of them. Consequently it has evolved many different weapon systems and responses against any type of invader. Discussing them all at once would be a bit much and make the already complex immune system even harder to understand. So for simplicity's sake, we will explain your complex defense mechanisms with the aid of your enemies. One at a time and one after another. Later on you'll get to know a few specific diseases and how they make your life miserable, and lastly we will look at internal dangers, such as cancer, allergies, and autoimmune diseases.

In part 2 of this book we will deal with some of the well-known microorganisms your immune system has to deal with: **Bacteria**. Bacteria are among the oldest living things on this planet and have been partying for billions of years. They are the smallest things we can consider alive without getting a headache. If, as we imagined earlier, a cell were the size of a human, the average bacteria would be the size of a bunny. Just like your own cells, bacteria are single-celled protein robots that come in a wide variety of shapes and sizes and are guided by chemistry and their genetic code. A common misconception about them is to think of them as primitive simply because they are smaller and less complex than our own cells.

But bacteria have been evolving for a long time and are exactly as complex as they need to be. And they are super successful on Earth! Bacteria are masters of survival and can be found basically everywhere where nutrients are found. And where none can be found, they sometimes just start making their own by finding ways to eat radiation or other formerly indigestible things. Bacteria saturate the soil you walk on, the surface of your desk, they float around in the air. They are on the page of the book you are reading right now. Some colonize the most hostile environments, like hydrothermal vents thousands of feet under the surface of the ocean, while others take on more pleasant places, like your eyelids.

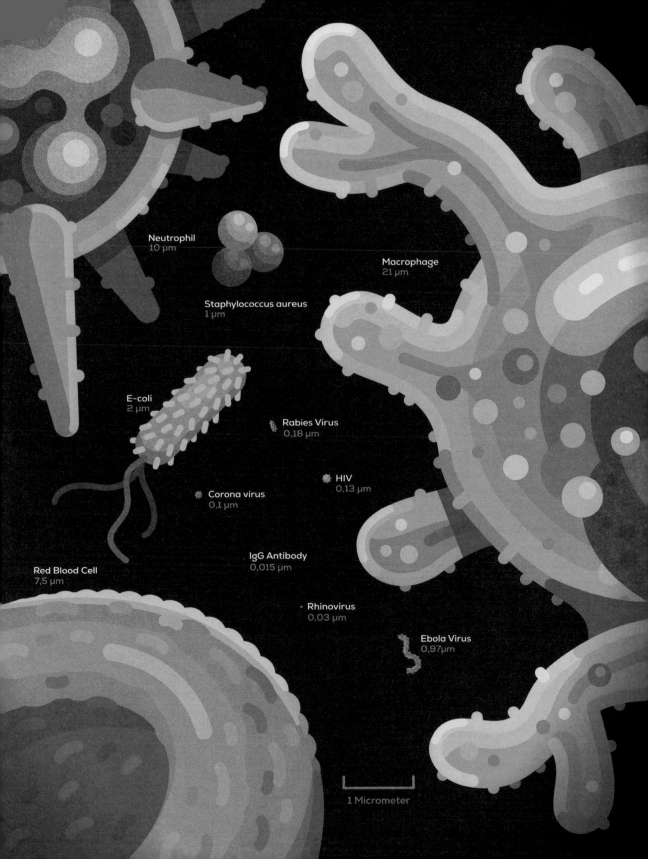

Neutrophil
10 μm

Macrophage
21 μm

Staphylococcus aureus
1 μm

E-coli
2 μm

Rabies Virus
0,18 μm

HIV
0,13 μm

Corona virus
0,1 μm

IgG Antibody
0,015 μm

Red Blood Cell
7,5 μm

Rhinovirus
0,03 μm

Ebola Virus
0,97μm

1 Micrometer

There has been some controversy about how large the combined bio-mass of all bacteria on earth is, but according to even the most conservative estimate, bacteria have at least ten times more mass than all animals combined. In one gram of soil, there are up to fifty million bacteria doing their thing. In one gram of the plaque on your teeth, more bacteria are living their life than there are humans on planet Earth right now (if you need a motivating story for your kids why they should brush their teeth and also give them nightmares, there you go).

In a pleasant environment, a single bacterium can reproduce once every twenty to thirty minutes by dividing into two bacteria. So after four more hours of dividing, there would already be 8,000 of them. A few more hours and there would be millions. And in a few more days, there would be enough bacteria to fill up the entirety of the world's oceans. Luckily this math does not quite work in reality because there is neither the space nor the nutrients. And not all species of bacteria can replicate this quickly, but this is what would technically be possible.

The point being, their potentially superfast reproductive cycle is a huge challenge for your immune system to deal with. Since they are so omnipresent on this planet, you are positively, absolutely covered by bacteria at all times, and have not even the slightest chance of ever getting rid of them. So our bodies had to arrange with this fact of life and make the best of it. A life without bacteria is impossible. And indeed, most bacteria are not only harmless to us, but our ancestors made a pretty good deal with them that is actually even beneficial to us. Trillions of them act as friendly neighbors and partners in crime and help you survive by keeping unfriendly bacteria away and breaking down certain food parts for you, and in return, they get a place to call home and free food. But these are not the bacteria we are concerned about in this book.

There are a lot of unfriendly, pathogenic bacteria that try to invade your body and make you sick. They cause a wide and scary variety of diseases, from diarrhea and all sorts of gut unpleasantries to tuberculosis or pneumonia or really scary things like the black plague, leprosy, or syphilis. If they get the chance, they also use any opportunity to infect your flesh when you injure yourself and get your insides in touch with the environment, where they

E-Coli Bacterium:

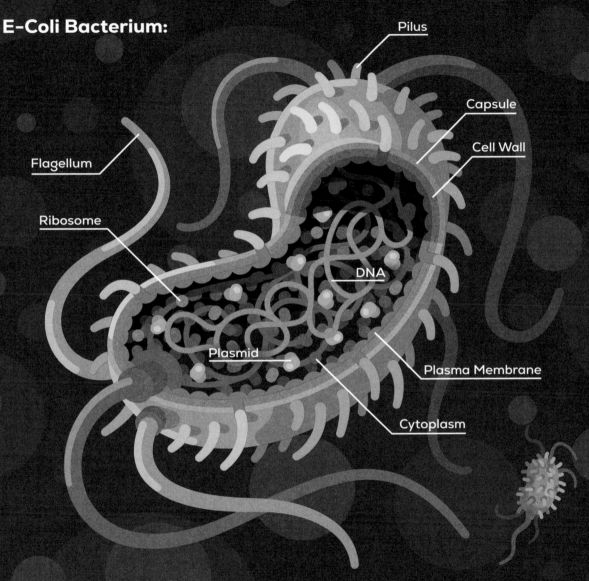

Pilus

Capsule

Cell Wall

Flagellum

Ribosome

DNA

Plasmid

Plasma Membrane

Cytoplasm

Bacterial Morphology:

Cocci

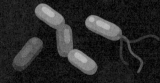

Rods

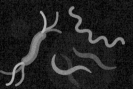

Spirals

exist just everywhere. Before the onset of antibiotics, even small wounds could lead to serious disease or even death.*

Even today with all the magic of modern medicine, bacterial infections are responsible for a good chunk of deaths each year. In other words, they are the perfect starting point to get to know your immune system! Let us see what happens when a few bacteria successfully make it into your body! But to get there, they first need to overcome a mighty barrier: The Desert Kingdom of the Skin.

---

* Let us give this throwaway line a bit more weight and remind all of us that our grandparents in fact did have it harder in life. We have data from a Boston hospital from 1941 that shows that 82% of bacterial infections of the blood resulted in death. We can barely imagine the horror this number represents—a scratch and a tiny bit of dirt literally could mean that your life was about to end. Today in developed countries less than 1% of these kinds of infections are deadly. The fact that we don't really think about this stuff too much shows how fast humans forget and move on, and how happy we can be to live in the present and not in the past.

6 The Desert Kingdom of the Skin

Your skin is the envelope of your insides, covering almost all the parts that you perceive as your outsides. It has the most direct contact with the world of any part of your body. This makes it crucial for the skin to be a really good border wall to protect you against all sorts of microbes trying to get in. Not only that, just through the process of living, it constantly gets damaged and hurt, so it needs to regenerate constantly. Luckily, the Desert Kingdom of the Skin is really good at all of that! There are a number of ingenious strategies the kingdom uses to be almost impossible to overcome for an intruder. The first one is that it is constantly dying. You can imagine your skin less as a wall and more like a conveyor belt of death. To understand how, we need to dive to the bottom, where your skin is created and produced.

The life of your skin cells begins around one millimeter deep. Here, the Skin Industrial Complex is situated. In the basal layer, stem cells do nothing but calmly multiply. They are cloning themselves, all day and night, producing new cells that begin a journey from the inside to the outside. The cells born here are special because they have a hard job. To be literally tough, not just figuratively, your skin cells produce a lot of keratin—a very tough protein that makes up the hard part of your skin, nails, and hairs. So your skin cells are hardy fellows, filled with special material that makes them hard to break.

As soon as they are born, they need to leave home. The skin stem cells constantly make new skin cells, and each new generation pushes the older ones further up. So your skin cells are constantly pushed upward by younger skin cells emerging below them. The closer they get to the surface, the more they need to become ready to be living defenders. And so as your skin cells mature, they develop long spikes and interlock with the other cells around them to form a dense and impassable wall. Next your skin cells begin manufacturing lamellar bodies, tiny bags that squirt out fat to create a waterproof and impermeable coat that covers the cells and the little bit of space that is left between them.

This coat does three things: It acts as another physical border that is extremely hard to pass; it makes it easier to dispose of dead skin cells later on; and it is filled with natural antibiotics called *defensins,* which can straight up murder enemies on their own. Your skin goes from newborn baby cell to expertly trained defender in the epic journey of a single millimeter.*

As the skin cells are pushed up further towards the surface they begin preparing for their final job: Dying. They become flatter and bigger and begin to stick together even tighter until they merge together into inseparable clumps. And then they shed their water and kill themselves.

Cells killing themselves is nothing special in your body, every second at least one million of your body cells go through some form of controlled suicide. And usually, when cells kill themselves, they do so in a way that's easy to clean up their corpses. But in the case of your skin cells, their dead bodies are actually very useful. You could even say that the purpose of their life is to die in the right place and become neat carcasses. The wall of merged, dead skin corpses is consistently pushed upward. Up to fifty layers of dead cells, fused together on top of each other, form the dead part of your skin that ideally covers your whole body.

When you look at yourself in the mirror, what you are really seeing is a very thin film of death covering your alive parts. As this dead layer of your skin is damaged and used up by the process of you living your life, it is constantly shed and replaced by new cells moving up from the stem cells deep below. Depending on your age, it takes your skin between thirty and fifty days to completely turn over. Every single second, you shed around 40,000 dead skin cells. So your outer border wall is constantly producing, emerging,

---

* Defensins are a really interesting beast. There are several subclasses and they are mostly produced by the border cells of your body and certain immune cells in battle. So what do they do? Well, they rip tiny holes into things. Think of them as tiny needles specific to certain intruders, like bacteria or fungi. So if these needles happen to encounter a microorganism they inject themselves into it and create a pore. A small wound, where the victim sort of bleeds a little bit. One needle will not kill a bacteria, but a few dozen might. Because the defensins are so specific, they are completely harmless for body cells but can kill microorganisms completely by themselves.

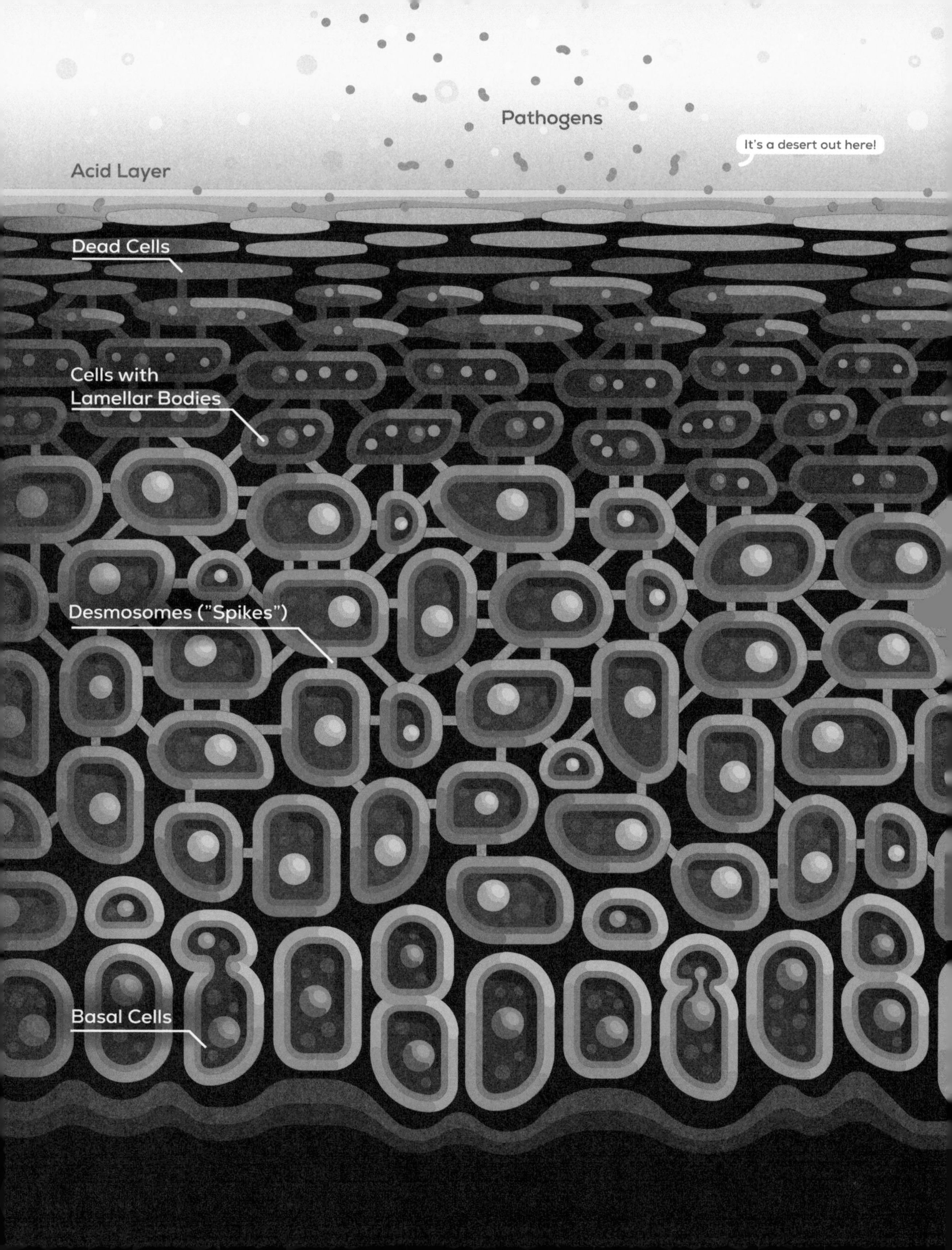

Pathogens
It's a desert out here!
Acid Layer
Dead Cells
Cells with
Lamellar Bodies
Desmosomes ("Spikes")
Basal Cells

and then discarding. Think about how ingenious and amazing a defense this is. Not only are the walls of the Skin Border Kingdom consistently replaced and fixed, as they move up they get coated in a fatty layer of passive and natural antibiotics. And even if enemies find a place to make their home and start eating away at the dead skin cells, they are consistently shed away from the body, making it much harder to gain a foothold on the skin.*

When it is warm, humans sweat a lot, which both cools us and also transports a lot of salt to the surface. Most of it is reabsorbed but some of it remains, overall making your skin a pretty salty place, which many microbes don't like. As if that's not enough, sweat contains even more natural antibiotics that can passively kill microbes.

So your skin does everything it can to be a real hellhole. From the perspective of a bacteria, it is a dry and salty desert filled with geysers that spit out toxic fluid and flush enemies away.

But this is still not all. Another one of the great passive defenses of your skin is that it is covered in a very fine film of acid, appropriately called the *acid mantle*, which is a mixture of sweat and other substances secreted by glands below your skin. The acid mantle is not so harsh that it would hurt you, it just means that the pH of your skin is slightly low and therefore slightly acidic and that is something a lot of microorganisms don't like. Imagine that your bed were sprinkled with battery acid. You would likely survive the night but you would suffer chemical burns and you would not be happy about your situation, and that's exactly how the bacteria feel.†‡

---

* We'll talk about viruses in a lot more detail in part 3 of the book, but since we are already here, we should mention that the way the skin is built, it is practically immune against viruses. Because these little parasites can infect only living cells and the surface of your skin consists only of dead cells, there is nothing to infect here! Only very few viruses have evolved ways to infect your skin, so bacteria and fungi are of much greater concern to your skin.

† pH—Acids and Bases: pH is one of those things that is often not properly explained or quickly forgotten once it is. For once, scientists gave something a great name: pH is short for POWER OF HYDROGEN, which is exciting and easy to remember. But then scientists decided to abbreviate it. Disappointing to say the least. Without getting too deep into it, the POWER OF HYDROGEN is a scale that describes how many hydrogen ions are present in a water-based solution.

‡ Wait, a footnote inside a footnote? Is this even allowed? Just to expand on the concept of "power." Power in this context doesn't mean that the hydrogen is extra strong or anything like

The acid mantle has another great passive effect mostly geared towards bacteria: The inside and outside of your body have different pH levels. So if a bacterium adapts to the acidic environment on the skin and then gets an opportunity to enter the bloodstream, for example, through an open wound, it has a problem: Your blood has a higher pH. So the bacterium suddenly finds itself in an environment that it is not adapted to, with very little time to do so, which is a considerable challenge to some species.

OK. So the skin is like a desert covered in acid, salt, and defensins and the ground is a graveyard of dead cells that are constantly shed away along with everybody unfortunate enough to sit on it. Learning about all of this, one might think that it is impossible for microbes to live on your skin. But this is far from the truth. In the endless universe of the microworld there is nothing like an uninhabited space. Everything is free real estate, no matter how hostile it is. But your body found a way to take advantage of this fact and make its defenses even tighter. Aside from your gut, which is basically made for and ruled by bacteria your body has invited in, your skin is the second most populated place of your body in terms of guests that are not you, but are indeed welcome. A healthy individual's skin holds up to forty different bacteria species, as different areas of your skin are drastically different environments with their own specific climates and temperatures. Your armpits, hands, face, and buttocks are pretty different places and house different guests. Overall, an average square centimeter of your skin is home to around a million bacteria. About ten billion friendly bacteria in total populate your

---

that. We are dipping into the wonderful universe of math here. This is "math power," correctly called exponential. So going up 1 POWER on the pH scale means having ten times fewer hydrogen ions. Going up 6 POWER on the pH scale means having one million times fewer hydrogen ions. (Why is going up the scale meaning fewer ions? Because the scale is inverted—why make something easy when it can be complicated?)

A lot of hydrogen ions means that something is acidic: Think of a tasty lemon or not as tasty battery acid. A low number of hydrogen ions means something is basic or alkaline, for example soap or bleach, both not very tasty. In general you don't want too many or too little hydrogen ions in a fluid because they will either take or donate protons. That's fine in weak acids, like when you squeeze a lemon over your food to make it taste better, but if a substance is either too basic or acidic it will act corrosively on your body. Corrosion means that it will destroy and decompose the structures your cells are made of and cause chemical burns. Small differences in POWER OF HYDROGEN make a bigger difference in the world of microbes.

outsides right now. And while you might not like this thought, you need them!

You can imagine these bacteria as a sort of barbarian horde in front of the gates. Your body has built a huge border wall and invited barbaric tribes to settle in front of it. They get to live off the land and enjoy free resources and space if they respect the border. As long as the balance is upheld, the Border Kingdom and the tribes live not only in harmony, but even in symbiosis. Should the barbarians try to enter though, maybe because the border is breached by an injury, the soldiers of the immune system will attack and kill them without mercy. So what do these billions of barbarian bacteria cells do for you? The most important thing is just taking up space. Squatting in a house is much harder if there are people already living there.

Your skin microbiome is pretty happy with its environment and does not intend to share it with strangers. So not only do they consume the available resources and physically occupy space, they communicate, regulate, and interact directly with the Border Kingdom and the immune cells living on the other side. For example, some of your bacterial guardians can produce substances that harm unwelcome guests. Going even further, it seems that they are even able to regulate the immune cells below the skin and tell it which harmful substances they should produce and in what quantities.

Once you reach adulthood, the composition of the microbes on your skin will remain relatively stable for the rest of your life, which means that there really is a shared benefit for your barbarian tribes and your body to find a balance and live in peace. It's an arrangement that everybody wants to uphold. Scientists don't completely understand yet how this deal is made, how the immune system decides who is allowed to settle, or how the bacteria educate the immune system about their intentions. But we know that this relationship exists and that it is very important.

Despite all those amazing defenses, the kingdom can be breached. Skin cells may be tough, but the world is tougher. And there are always bacteria ready to take a chance if they get one. Let us witness the immune system in real action for the first time.

Before we dive into a story, a short note: The way we will describe an infection and the response of the immune system is an idealized example. One where things happen in a clear sequence, one level of escalation after another, each level clearly triggered by the one that came before. So just keep in mind—reality is more complex. We are simplifying without dropping too many details and organizing things in a nice straightforward manner. OK, having that out of the way, let us destroy your skin and challenge your immune system!

7 The Cut

SMALL ACTIONS CAN HAVE BIG CONSEQUENCES. SIMPLE MISTAKES CAN LEAD to catastrophic outcomes. Something mildly annoying on the scale of the human giant is a full-blown emergency on the scale of your cells.

Imagine strolling through the woods on a pleasant summer day. It's hot and humid and you went with light and fashionable shoes instead of your nature-proof boots because this is the woods, not the jungle, and you are an adult and can make your own decisions! You hike up a hill when you suddenly feel a sharp pain. You look down and see that you stepped on a rotten board that once was nailed to a tree but has since decided to become a death trap. A long and rusty nail penetrates the sole of your shoe. You pull it out and curse and complain a bunch about the world in general and your cruel fate in particular. Nobody could have seen this coming. It doesn't hurt too bad though. You take your shoe and sock off to look and it's nothing terribly serious, only a little bit of bleeding. So you continue hiking and muttering under your breath.

Meanwhile your cells had a pretty different experience. When the nail penetrated your shoe, its tip entered your big toe. It ripped through your skin like a pointy piece of metal would. For your cells, it was an average day until suddenly, their world exploded. From their perspective a large metal asteroid just ripped a hole into the world. Much worse though, it was covered with soil and dirt and hundreds of thousands of bacteria that suddenly found themselves beyond the gates of your otherwise impenetrable skin border wall. And now this has become a whole situation.

Immediately, bacteria spread into the warm caverns between helpless cells, ready to consume nutrients and explore a little. This here is much better than soil! There is food and water and it is warm and comfy and there are only victims around that seem like helpless children. The bacteria have no

48

intention of leaving ever again. And the soil bacteria are not the only unwelcome visitors. Thousands of bacteria that did their thing on the surface of your skin and in your moist socks now also decide to check out this paradise that has just appeared out of nowhere. What a great and lucky day!

Your body politely disagrees with this assessment. Hundreds of thousands of civilian cells have died, ripped apart by the foreign object that so suddenly crashed through the sky. Others are wounded and distressed. And similar to a catastrophe on the human scale, the civilians are screaming in terror, sending messages of alarm and panic to everyone ready to receive them. These panic signals, the guts of dead cells, and the stench of thousands of bacteria are carried into the surrounding tissue, raising an urgent alarm.

Your Innate Immune System reacts immediately. Sentinel cells are the very first to show up—they were peacefully patrolling the premises when the impact happened and are quickly making their way right to ground zero, attracted by the screams and the detritus of the crash site. These cells are called **Macrophages** and they are the largest immune cells your body has to offer. Physically, Macrophages are pretty impressive. If an average cell would be the size of a human, a Macrophage would be the size of a black rhino. And just as with black rhinos, you better not mess with them. Their purpose is to devour dead cells and living enemies, coordinate defenses, and help heal wounds. Jobs that are in high demand, because right now determined bacteria are proliferating rapidly—they need to be stopped quickly before they can establish a real presence.

The chaos puts the Macrophages into a rage they have never experienced before. Within seconds, they engage the bacteria in battle and throw their own bodies violently at them—imagine a wild rhino trying to stamp panicked bunnies to death. But the bunnies, obviously, prefer not to be stamped to death and so they flee and try to escape the grasp of this powerful cell. But their escape plan will be in vain, as Macrophages are able to stretch out parts of themselves, a bit like the arms of an octopus, guided only by the smell of the panicked bacteria. When they manage to grab one of them, its fate is sealed. The grip of a Macrophage is too strong, and resistance is futile, as it pulls the unlucky bacteria in and swallows it whole to digest it alive.

But despite the cruel efficiency and forceful effort, the wound is too mas-

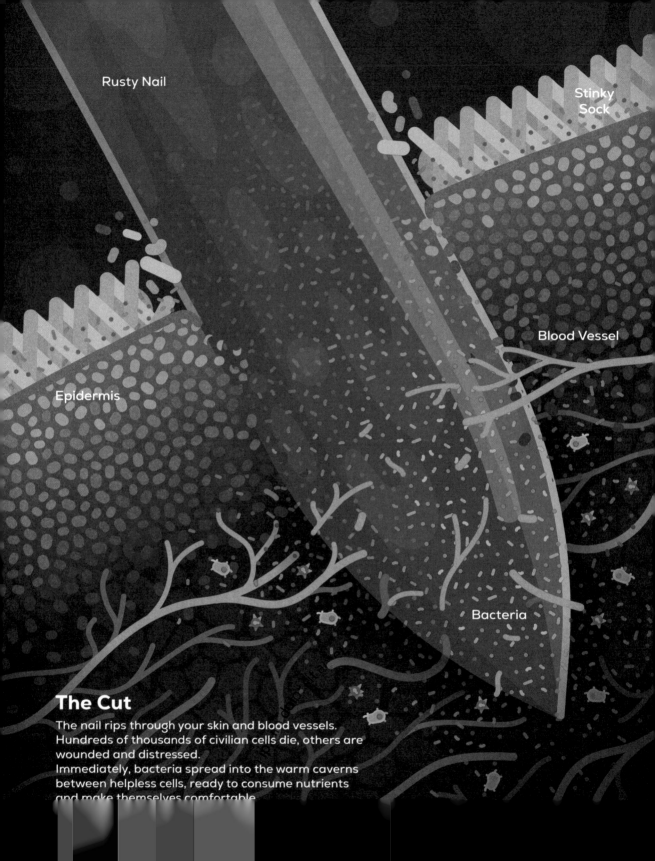

Rusty Nail

Stinky Sock

Blood Vessel

Epidermis

Bacteria

The Cut

The nail rips through your skin and blood vessels.
Hundreds of thousands of civilian cells die, others are
wounded and distressed.
Immediately, bacteria spread into the warm caverns
between helpless cells, ready to consume nutrients
and make themselves comfortable.

sive, the damage too great, the exposed surface too big. As the Macrophages devour one enemy after another they realize that they can at best slow this invasion down, not stop it. And so they begin to call for help, sending out urgent alarm signals, and start preparing the battlefield for reinforcements that will arrive shortly. Lucky for them, backup is already under way. In the blood thousands of **Neutrophils** have heard the cries for help and smelled the signs of death and begun to move. At the site of infection they leave the rushing ocean of the blood and enter the battlefield. Just like the Macrophages, the panic and alarm signals activated them, turning them from pretty chill fellows into killing frenzied maniacs.

Immediately they begin hunting and devouring bacteria whole, but with much less care for their surroundings. Neutrophils are on a tight timer: Once active they only have hours before they will die of exhaustion as their weapons do not regenerate. So they make the best of the situation and use them freely—not only killing enemies but also causing real damage to the tissue they should be protecting in principle. But collateral damage is not their concern at this moment or ever, as the danger of bacteria spreading through the body is much too grave to consider civilians. But they do not only fight, they also self-sacrifice—some of them explode, casting wide and toxic nets around themselves in the process. These nets are spiked with dangerous chemicals that seal off the battlefield, trap and kill bacteria, and make it harder for them to leave and hide.

Back in the world of the humans, you sit down again to take a second look at the damage. The small wound is already covered by a very thin film of crust. At this point the wound has already closed superficially as millions of specialized cells from the blood flooded in the battlefield: Platelets, blood cells that exist mainly to act as an emergency worker that closes wounds. They produce a sort of large, sticky net that clumps themselves and unlucky red blood cells together and creates an emergency barrier to the outside world, stopping blood loss rather quickly and preventing more intruders from entering. This enables fresh skin cells to slowly start closing the enormous hole in the world.*

---

* OK get ready for a wild story! Platelets are not actually cells but cell fragments from another cell called megakaryocyte. Enormously large fellows, that are around six times larger than your aver-

Overall your toe has swollen a bit and feels warm and hurts slightly. Annoying for sure, but no big deal you think as you curse your carelessness and get ready to continue your walk, with a slight limp. Or so you think. What you experience as a light swelling is a purposeful reaction of your immune system. The cells fighting at the site of infection started a crucial defense process: **Inflammation**.

This means they ordered your blood vessels to open up and let warm fluid stream into the battlefield, like a dam opening up towards a valley. This does a few things: For one, it stimulates and squeezes nerve cells that are deeply unhappy about their situation and send pain signals to the brain, which makes the human aware that something is wrong and an injury occurred.

Still, that does not help with the hundreds of thousands of enemies that already made their way in, but luckily the rush of fluid caused by the inflammation carries a silent killer into the battle zone. Many bacteria get stunned, or begin twitching as dozens of tiny wounds mysteriously appear on their surfaces and make them ooze out their insides, which is pretty bad and kills them. We'll get to know this silent killer intimately at a later point.

As the furious fighting rages on and as more and more bacteria are killed, the first immune soldiers are dying too. They have given all they got and now they just want to sleep. Millions of soldier cells continue to stream in and bash in as many heads as they can before they die themselves. We reach a crossroads. The battle can go multiple ways now. In most cases, if things go well, this will more or less be the extent of the damage. All bacteria are killed and the immune system assists the civilian cells to heal. In the end

---

age cell and live in your bone marrow. They have very long octopus-like arms, that they push into your blood vessels and begin to grow them. When one of these weird arms has grown enough, small packages break off. Functional mini parts of cells that are carried away by the blood. These packages are your platelets and every time you cut yourself or suffer a wound, they close it. A single megakaryocyte produces around 10,000 of these platelets in its lifetime from its flabby long arms that stretch from your bones to into your blood. Your body is seriously so weird and so amazing.

this turns out to be a tiny wound, the kind you sustain all the time and never think about.

But in this story things do not go well. Among the intruders there is a pathogen. A soil bacteria that is actually able to deal with the immune response and to multiply quickly. Bacteria are living things and able to react to situations. And so they do, setting off defense mechanisms that make them harder to kill or resistant against the weapons of the immune system. The best the Innate Immune System can do is to keep them in check.

So another immune cell now makes a serious decision. It has been acting quietly in the background, monitoring the events at the battlefield as they were unfolding. Now, hours after the catastrophe happened and the infection began, its time to shine has finally come.

The **Dendritic Cell**, the mighty messenger and intelligence officer of the Innate Immune System did not just watch the disaster unfold. Dendritic Cells are stationed everywhere the Border Kingdom can be penetrated. Shaken up by the chaos and panic, they urgently began collecting samples of the battlefield. Similarly to Macrophages, Dendritic Cells have long tentacles to catch invaders and rip them into pieces. But their goal was not to devour them—no, they prepared samples made from the dead intruders, to present their findings to the intelligence centers of the immune system. After a few hours of sampling, the Dendritic Cells get on the move, leaving the battlefield behind to get help from the Adaptive Immune System. It takes the Dendritic Cell about a day to reach its destination and when it finds what, or better, who it is looking for, a beast will rouse from its sleep and all hell will break loose.

Let us pause here for a moment and consider how well your body was prepared for this emergency. Cuts and bruises and stings by rusty and pointy things are nothing we really worry about. It is just a fact of life that you hurt from time to time and it almost never goes beyond being slightly annoying. If the infection can't be stopped a course of antibiotics usually does the trick but for the vast, vast majority of human history such powerful medication was not available and a small injury could actually be deadly.

So your body had to evolve ways to brutally and quickly crush an invasion that inevitably followed when your border kingdom got compromised. And oh boy, is your Innate Immune System good at that. We only very briefly met the cells who are your first line of defense, the Macrophage, the Neutrophil, and the Dendritic Cell, but actually they can do much more! And what about the mysterious invisible force that killed and stunned the invaders but that we did neither name nor describe.

8 The Soldiers of the Innate Immune System: Macrophages and Neutrophils

As we were just able to witness, Macrophages and Neutrophils are the damage dealers of the Immune system. Together they are a special class of cells called *phagocytes*. Not the worst of all names in immunology actually, as it means "Eating Cell." And eat they do. *Macrophage* means Great Eater, which is a great fit. Since cells don't have tiny mouths, eating on this level has to mean something else.

Imagine you had no mouth and wanted to eat like a phagocyte, it would look something like this: You take a sandwich and hold it against your skin. Where is not important, anyplace on your body will do. Your skin folds in on itself and pulls the sandwich into your insides, trapping it in a bag of skin that now floats to your stomach and merges with it, dropping the sandwich in your stomach acid.

Disturbing in the human world, very practical in the cellular world. The process is actually pretty fascinating. When a phagocyte, like a Macrophage, wants to swallow an enemy, it reaches out to it and grabs it tightly. Once it has a firm grip, it pulls its victim in, folds a part of its membrane into itself, and engulfs the victim, trapping it in a sort of mini prison that is now inside the Macrophage. In a sense, a part of the outside of the Macrophage becomes a sort of tightly closed garbage bag that is pulled inside. The Macrophage is equipped with an abundance of compartments that are filled with the equivalent of stomach acid—substances that dissolve things. These compartments then merge with the tiny prison and pour their deadly contents all over the victim, dissolving it into its components, into amino acids, sugars, and fats that are not only harmless, but even useful. Some become

Phagocytosis

1. Phagocyte grabs on to pathogen.

2. Folding its membrane in, it traps the enemy in a mini prison.

3. The mini prison merges with a compartment filled with acid.

4. The acid breaks the pathogen down into basic components.

5. The phagocyte eats the parts it can and spits out the rest for other cells to eat.

food for the Macrophage itself and others are spat out so other cells can have a meal too. Life hates nothing as much as wasting resources.

This process is extremely important because it is the main way your body gets rid of whole armies of invaders and their garbage. Indeed, one of the main jobs of Macrophages is to eat and swallow stuff that the body does not want around, battle or not.

Interestingly, the main things Macrophages eat are actually parts of you. Most cells of your body are on a limited life timer to avoid becoming faulty and turning into something bad, like cancer, for example. So every second of your life, around one million of your cells die by controlled cell suicide, called *apoptosis* (this process will come up a few times in this book because it is very important). When cells decide their time has come, they release a special signal letting everybody else know that they are done. Then they destroy themselves via apoptosis, which means that they split up into a bunch of small, neat packages of cell garbage. Macrophages, attracted by the signals, pick up the shreds of the former cells and recycle the parts.

Macrophages probably are an extremely old invention of the immune system, maybe even the first sort of dedicated defense cell, since almost all multicellular animals have some form of macrophage-like cell. In a sense they are a bit like single-celled organisms. Their main job is border patrol and garbage disposal unit, but they also help with coordination of other cells, preparing the battlefield by causing inflammation, and encouraging wound healing after an injury. As a bonus they did not ask for, if you have a tattoo, a lot of it is probably stored in your Macrophages.*

---

* Have you ever asked yourself why your body would be OK with having huge amounts of ink below its skin? Because in general, your immune system is not OK with anything that is not itself, or did not get a special permit, hanging around in the body. But somehow, you can push ink with a fast-moving needle into the second layer of your skin and it remains there for many years. While the body is not excited about having ink under its skin, if the person scratching tasteful art into your flesh is doing their job correctly, it is also not particularly harmful. Still, your local immune system is not pleased by the intrusion. And so your skin swells and some of the ink particles are carried away. The majority stays in the tissue though—not because Macrophages do not try to gobble them up. Most of the metallic ink particles are just too large to be swallowed, and so they remain where they are. The ones that are small enough to be eaten, though, are eaten.

While Macrophages are really great at breaking down bacteria and other cell garbage, they aren't actually able to destroy the ink. So they just keep it inside themselves and store it. If you have a tattoo, when you look at it remember that it is partially trapped within your immune sys-

Macrophages live up to several months. Billions of them hang out just below your skin, patrolling the surfaces of things like your lungs and the tissue that surrounds your intestines. Billions more hang out throughout your body. In your liver and spleen, they catch old blood cells and eat them whole to recycle the valuable iron they carry. In your brain they make up around 15% of all cells and are extra calm, so they don't accidentally damage irreplaceable nerve cells that you need for important things like thinking about movies or breathing air.

Macrophages don't have very exciting lives. They hang around in the area they are responsible for, wobble around, and pick up garbage and dead cells. But if they become annoyed, they become terribly scary fighters. An activated angry Macrophage can swallow up to 100 bacteria before it dies of exhaustion. For a long time it was assumed that this was what they were, a sort of aggressive janitor—but it turned out Macrophages actually play many different roles and interact with a lot of different cells for a variety of jobs.

So it might be better to look at Macrophages as a sort of local captain of the Innate Immune System: In battle they tell other cells what to do and inform them if fighting is still necessary.

Lastly, when an infection has been dealt with, Macrophages actually can slow or even shut down the immune response at the site of battle to prevent further damage. An ongoing immune reaction is not good for you because immune cells generally put stress on the body and waste a lot of energy and resources. So as a battle dies down, some Macrophages turn a battlefield into

---

tem. Unfortunately if a few years later you decide that the Chinese characters that turned out to mean "soup" are not as tasteful anymore and want to get them removed, your immune system also makes it really hard to get rid of a tattoo.

The most common process of tattoo removal is a special laser that permeates your skin and heats up the ink particles so much on one side that it is put under tension and breaks apart into smaller pieces. Some float away, others are now eaten by Macrophages. This can make tattoo removal very hard, because even though old ink-filled Macrophages die at some point, young replacements arrive and swallow the remains of their dead predecessors, complete with all the ink inside. Again they can't destroy it and so they just store and ignore it. This way tattoos stay visible for years. Over time, with each new cycle of replacement some of the ink gets lost and swept away in the process, or a few of the new Macrophages move a tiny bit around. So your tattoo will fade out and becomes less sharp on the edges.

a friendly construction site and literally begin to eat the remaining soldiers. Then they release chemicals that help the civilian cells regenerate and re-build damaged structures like blood vessels, so your wounds can heal faster. Again, the immune system hates to waste anything.

The **Neutrophil** is a bit of a simpler fellow. It exists to fight and to die for the collective. It is the crazy suicidal Spartan warrior of the immune system. Or if you want to stay in the animal kingdom, a chimp on coke with a bad temper and a machine gun. A sort of all-purpose weapon system that is en-gineered to quickly deal with the most common enemies your body encounters—especially bacteria. It is by far the most abundant immune cell in your blood and easily one of the most potent. Neutrophils are indeed so dangerous that they come with a kill switch. They are on a tight timer and only live a few days when they are not needed before they commit controlled suicide.

But even in battle their life is short and lasts only a few hours. The risk of them wreaking havoc on the infrastructure of your body is simply too high. And so every single day 100 billion Neutrophils give up their lives voluntarily and die. And every single day around 100 billion more are born, ready to fight for you if necessary.*

But despite the danger they present to your body, they are indispensable for your day-to-day survival and without them your defense would be seri-ously maimed. When in combat, Neutrophils have two additional weapon systems on top of eating enemies alive. They can throw acid at enemies and kill themselves to create deadly traps. Neutrophils are densely packed with *granules*, which are basically tiny packages filled with a deadly load. You can imagine these granules as little knives and scissors that are made to cut open and cripple intruders. So if a Neutrophil encounters a bunch of bacte-ria in one place it will just shower them with granules that rip their outsides apart. The problem with this approach is that it is not super specific, and it hits whoever is unlucky enough to be in the way. This often means your own healthy civilian cells. And this is one of the reasons the body is sort of afraid

---

* In actuality, you produce around one billion Neutrophils per two pounds (a kilogram) of body weight; you can do the math for what this means for you.

of Neutrophils. They kill very efficiently, but they can potentially cause more harm than good if they get too excited.*

But the most mind-boggling thing Neutrophils do in battle is to create deadly nets of DNA, sacrificing themselves in the process. To get an idea what this means, imagine you were a burglar and wanted to break into a museum at night to steal and let your buddies in to have a stealing-stuff party. So you are doing a great job and sneak by cameras and security systems, entering the vaults where all the valuables are. "Things are going great," you think as you begin stuffing paintings into your backpack.

But then, suddenly you see a guard charging at you screaming—you get ready for a fight. But instead of swinging at you, the guard rips open his chest, splitting his ribs into countless sharp splinters while pulling out his intestines. You don't even have time to get confused before he starts swinging his guts, spiked with sharp bone splinters at you, like the world's most disgusting whip. You cry in pain and confusion as he mercilessly strikes you, causing deep wounds and leaving you stunned and unable to flee. And then he punches you in the face. "This did not go as expected," you think, as he begins eating you alive.

This is what Neutrophils do when they create a *Neutrophil Extracellular Trap*. Or *NET* for short. If Neutrophils get the impression that drastic measures are called for, they begin this crazy kind of suicide. First their nucleus begins to dissolve, freeing up their DNA. As it fills up the cell, countless proteins and enzymes attach to it—the sharp bone splinters from our little story. And then the Neutrophil literally spits out its entire DNA around itself, like a giant net. Not only can this net trap enemies in place and hurt them—it also creates a physical barrier that makes it harder for bacteria or viruses to

---

* Neutrophils are indeed so careless when it comes to causing collateral damage that there are cases where Macrophages are trying to hide damaged cells from them! Every day for a variety of reasons, a few cells die in your organs in unnatural ways, maybe because you looked at your phone and ran into a street sign for example. But often enough the damage is pretty mild and does not require a strong reaction from your immune system. We will learn more about it later, but dead cells attract Neutrophils and if they find a single dead cell, they will escalate the situation and cause even more damage that is not necessary. So to stop them from that, Macrophages can sort of cover a single dead cell with their body to literally hide it from Neutrophils, so they are confused and leave again.

escape and move deeper into the body. Usually the brave Neutrophil dies doing this, which seems obvious.

Sometimes, even though they vomited out their DNA, these brave warriors continue to fight, throwing acid at enemies or swallowing them whole and doing Neutrophil stuff before they finally die of exhaustion. The question could be asked if a cell that has given up its entire genetic material is still alive. In any case, it can only go on for so long—without DNA a cell has no way to maintain its inner machinery. Whatever this cell is—a living entity or no more than a zombie following its last commands mindlessly—it keeps doing what it was made to do: It fights and dies for you, so you can live. No matter which of its weapon systems it uses, the Neutrophil is one of your fiercest soldiers and one that enemies, and our own bodies, are rightfully pretty afraid of.*

Macrophages and Neutrophils have another important job that they share with other parts of the immune system, and that we will discuss in more detail in the next chapter because it is so utterly crucial for your defenses: They cause inflammation, a process so important to your defense and your health that we need to check out what is going on here. So before we can come back to the battlefield and your army defending you, we will do a short excursion and learn about a few fascinating and very important mechanisms your immune system uses in a fight.

---

* Another neat little detail about Neutrophils is that when they are chasing a pathogen, they often do so in swarms that follow the same mathematical rules as swarms of insects. So imagine being hunted by a bunch of hornets the size of cows and you get the same stressful experience that many bacteria must go through in the last few moments of their lives.

9 Inflammation: Playing with Fire

INFLAMMATION IS SOMETHING YOU PROBABLY NEVER HAVE THOUGHT about that much, since it's pretty banal. You hurt yourself and the wound swells and reddens a bit, big deal. Who cares. But actually, inflammation is critically important for your survival and your health, enabling your immune system to address sudden wounds and infections.

Inflammation is the universal response of your immune system to any sort of breach or damage or insult. No matter if you burn yourself, cut yourself, or get a bruise. No matter if bacteria or viruses infect your nose, lungs, or gut. No matter if a young tumor kills a few civilian cells by stealing their nutrients or you have an allergic reaction to a food, inflammation is the response. *Damage or danger—perceived or real—causes inflammation.*

Inflammation is the red swelling and itching from an insect bite, the sore throat when you have a cold.

In a nutshell, its purpose is to restrict an infection to an area and stop it from spreading, but also to help remove damaged and dead tissue and to serve as a sort of expressway for your immune cells and attack proteins directly to the site of infection!*

Paradoxically, inflammation is also one of the most unhealthy things that can happen to you if it becomes chronic. According to pretty new science,

---

* The way inflammation helps immune cells to get to the battlefield is very weird and fascinating. What basically happens is that the chemical signals from inflammation trigger a change in your blood vessels close to where the inflammation signals are coming from, and in your immune cells that get activated by these signals. Both parties extend many little special adhesion molecules that work a bit like Velcro. The immune cells that are speeding through the blood now can stick to the cells making up the blood vessels, and slow down close to the site of infection. On top of that, inflammation makes your blood vessels more porous, which makes it easier for your immune cells to squeeze through tiny spaces and move towards the battlefield.

chronic inflammation is involved in more than half of all deaths each year as it is an underlying cause of a wide variety of diseases—from various cancers to strokes or liver failure. And yes, you read that correctly—*for at least one in two people who died today, chronic inflammation was the underlying cause of the disease that killed them.* Despite the fact that chronic inflammation is so taxing to the body, "regular" inflammation is indispensable for your defense.

Inflammation is a team effort, a complex biological response of your immune system to mount a rapid defense against injury or infections. In a nutshell, inflammation is a process that makes the cells in blood vessels change their shape, so that plasma, the liquid part from your blood, can flood into a wounded or infected tissue. You can literally imagine this as floodgates opening and a tsunami of water, filled with salts and all sorts of special attack proteins, flooding the spaces between your cells so rapidly that tissue on the scale of a metropolitan area balloons up. Wherever your cells suspect that something fishy is going on they order inflammation as a dramatic first response.*

You can tell if you have inflammation through five markers: Redness, heat, swelling, pain, and loss of function. So for example, in our story where you stepped on a sharp nail, the injured toe was flooded with fluid and swelled up, so it became red because of extra blood in the tissue.

The injured toe becomes hot as the blood brings extra body heat. This heat does useful things for you: Most microorganisms do not like it hot—so making the wound hotter slows them down and makes their lives stressful. And you want pathogens in your body to be as stressed as possible. In contrast, your civilian repair cells like the extra temperature very much as it speeds up their metabolism and enables your wound to heal faster.

Then there is pain. Some of the chemicals released by inflammation make your nerve endings more susceptible to pain and through the process

---

* Everything related to the immune system has an exception. There are a few areas in your body that are excluded from this rule, like your brain, spinal cord, part of your eyes, and testicles (if you happen to have testicles, that is). All are extremely sensitive regions where inflammation could cause immediate, irreparable damage and so these areas are so-called immune privileged, which means that cells of the immune system are kept out of there by the blood-tissue barriers—and those that are allowed in are on super special orders to behave.

Inflammation

Inflammation is a complex biological response of your immune system to mount a rapid defense against injury or infection. Blood vessels flood the tissue resulting in redness, heat, swelling, pain, and loss of function.

Inflamed Injury

Macrophage

Neutrophil

Battlefield

Epidermis

Blood Vessels

Blood vessel releasing plasma, flooding the battlefield with fluids, proteins, and new soldiers.

of swelling, there is pressure on nerve cells with pain receptors, motivating them to send whiny signals to the brain. Pain is a very effective motivation in the sense that we prefer not feeling it.

And lastly there is loss of function. This one is pretty straightforward—if you burn your hand and inflammation makes it swell up and hurt, you can't use it properly. Same with stepping on a nail—your foot is just not happy about this event. Together with the pain, this loss of function makes sure that you rest and don't burden or strain your wounded body part. It forces you to give yourself time to heal. These are the five hallmarks of inflammation.

As we will encounter over and over again in this book, inflammation is very hard on the body as it causes stress in the affected tissue and brings in agitated immune cells like Neutrophils that do damage, so it has a few built-in mechanisms that damp it down again. For example, the chemical signals that cause inflammation are used up pretty quickly. So inflammation needs to be requested consistently by your immune cells or it just dissipates by itself. You might ask, but what exactly is causing inflammation? Well, a variety of mechanisms.

The first way inflammation gets started is through dying cells. Amazingly, your body evolved a way to recognize if a cell died a natural way or if it died a violent death. The immune system has to assume that cells dying an unnatural death means grave danger, and so death is a signal that causes inflammation.

Normally, when a cell has reached the end of its life, it kills itself through apoptosis that we encountered already. Apoptosis is basically a calm suicide that keeps the contents of the cells nice and tidy. But when cells die in unnatural ways, for example by being ripped into pieces by a sharp nail, burned to death by a hot pan, or poisoned by the waste products of a bacterial infection, the insides of your civilian cells spill all over the place. Certain parts of the guts of your cells, like DNA or RNA, are high-alert triggers for your immune system and cause rapid inflammation.*

---

* You probably learned in school that mitochondria, the powerhouse of the cell, were ancient bacteria that merged with the ancestors of your cells to become a symbiotic organism. Today they are organelles inside your cells that provide the cell with useful energy. Your immune system still

This is also a good moment to introduce a very special cell that you might learn to hate later on when we learn more about it—if you ever had a severe allergic reaction where your body did swell up explosively, this cell had most likely a hand in it: The **Mast Cell**. Mast Cells are large, bloated cells filled with tiny bombs containing extremely potent chemicals that cause rapid and massive local inflammation. (For example, the itching you feel when a mosquito bites you was probably caused by chemicals the Mast Cell released.) They mostly sit below your skin and do their thing, which is thankfully not much. If you hurt yourself and tissue is destroyed and they die or they get really agitated, Mast Cells release their inflammation supercharger chemicals and speed up the process immensely.

This way the tissue below your skin has an inflammation emergency button. This might be a good place to note that some immunologists believe that the Mast Cell plays a much more direct and central role in the immune system, although that is not part of most textbooks. The great thing about science is that proving established ideas wrong is a win for everybody, so we'll know in a few years if Mast Cells deserve more love.

The next best way to cause inflammation is more of an active decision: Macrophages and Neutrophils order inflammation when they are engaged in a battle. This way, as long as fighting is going on, they release chemicals that keep the battlefield flooded and ready to take in fresh reinforcements. But this is also one of the reasons why having any kind of battle going on for a long time is bad.

For example, if you have a lung infection like pneumonia or COVID-19, inflammation and the fluid that it summons into the lung tissue can make it hard to breathe and create feelings of drowning. The feeling is horrifyingly accurate in this case, as you are literally drowning in extra fluid, only fluid that came from the inside rather than the outside.

OK, enough about inflammation for now—just to summarize: If your cells are dying unnaturally, if you rupture or annoy a Mast Cell below the skin, or if your immune system is fighting enemies, they release chemicals

---

remembers them as bacteria though, as intruders that have no business being outside cells. So if your cells burst and your immune system detects mitochondria floating around, your immune cells will react super alarmed.

that cause inflammation. A flood of fluids and all sorts of chemicals, which annoy your enemies, attract reinforcements, and make it easier for them to get into the infected tissue, all of which make it easier to defend a battlefield. But inflammation is hard on the body and in many cases presents a real danger to the health of the body.

10 Naked, Blind, and Afraid: How Do Cells Know Where to Go?

AT THIS POINT WE HAVE TOTALLY BEEN IGNORING ANOTHER PRETTY IMPORtant detail: How do cells know which way is which and where to go and where they are needed? When we imagine cells as people and remember that they're patrolling the equivalent area of the European continent, one of the first questions you might have is, how could they possibly go the right way? Wouldn't they get lost constantly? Also, to make this a bit more challenging, cells are blind, which makes sense if you think about it for a moment.

The process of seeing something requires light waves to hit the surface of an object, bounce back from that surface, and hit a sensory organ like your eye, where a few hundred million specialized cells transform them into electrical signals that are sent off to your brain for interpretation. All of this seems to be a bit too much of an investment for single cells.*

Even if cells did have eyes, on their scale, "seeing" would not be very useful. Because their world is really, really small, and for a single cell, light waves are huge and impractical. If you were the size of a cell, the waves of visible light would reach from your toes to your belly button! Bacteria are already so small that they are barely visible with light microscopes and images of them are pretty grainy. And viruses are even smaller, considerably smaller than light waves and therefore invisible in any sense that we consider seeing, except with specialized tools like electron microscopes. Besides, most places inside your body are pretty dark. If your insides are well lit, something has gone horribly wrong.

---

* Yes, there are single-celled organisms that have photoreceptors, enabling them to note the difference between dark and not dark and the direction where light is coming from. But this is not what we are talking about here.

The same principle is true for hearing, which is the ability to detect changes in the pressure of gases and fluids and transform these differences into information. Another thing we have dedicated organs for and that fits the environment humans live in and is impractical for cells. OK, "seeing" and "hearing" in the sense humans are used to is not a great option in the microworld. So how do cells experience their world? How do they sense it and how do they communicate with each other?

Well, in a way, cells smell their way through life. For cells, information is a physical thing: **Cytokines**. In a nutshell, cytokines are very small proteins that are used to convey information. There are hundreds of different cytokines and they are important in almost every biological process going on inside of you—from your development in your mother's womb to the degeneration you experience as you get older. But the field where they are the most relevant and important is your immune system. They play a crucial role in the development of diseases and how your cells are able to respond. In a sense cytokines are the language of your immune cells. We will encounter them a few times in this book, so let us get an idea of what they do:

Let's say a Macrophage floats around and stumbles over an enemy. The discovery needs to be shared with other immune cell buddies, so it releases cytokines that carry the information *"Danger! Enemy around! Come help out!"* These cytokines then float away, carried purely by the random motion of particles in your bodily fluids. Somewhere else, another immune cell, maybe a Neutrophil, smells these cytokines up and "receives" the information. The more cytokines it picks up, the stronger it reacts to them.

So when the rusty nail penetrated our skin and caused untold death and destruction, thousands of cells screamed in unison and released a very high amount of panic cytokines, which translates into the information that something horrible had happened and they needed urgent help, alerting thousands of cells to move. But this is not all, the smell of cytokines also functions as a navigation system.*

* OK, technically we could be more precise here. There are two general classes of cytokines relevant at this point: The cytokines that convey information and chemokines. Chemokines are a family of small cytokines secreted by cells. Their name means "move chemical" and is very fitting as their main ability is to motivate cells to move in a certain direction. They are not just floating around, certain civilian cells also can pick them up and sort of "decorate" themselves with them,

Cytokines

Cytokines are very small proteins that are used to convey information. They play a crucial role in the development of diseases and how your cells are able to respond. In a sense cytokines are the language of your immune cells.

The closer to the origin of the source of a smell a cell is, the more cytokines it will pick up. By measuring the concentration of cytokines in the space around it, it can precisely locate where the message is coming from and then begin moving in that direction. It is sort of "smelling" where the smell is the most intense. Which will lead it to the site of battle.

To do this, your immune cells don't have a single nose, they have millions of them, all over their bodies, covering their membranes in every direction.

Why so many? For two reasons: First, by being covered by noses, cells have a 360° smell system. They can pretty precisely tell from which direction a cytokine is coming from. These noses are so sensitive that for some cells, as little a concentration difference as 1% in the cytokine signals around a cell is enough to tell it where to go. (Which is a fancy way of saying that there might just be 1% more molecules on one side of the cell.) This information is used to orient the cell in space and then make it move towards its target, always following the path where the most cytokines are coming from. A cell takes a step and then a whiff. And then it takes another step and takes another whiff. Until it gets to where it is needed.

The other reason it is good to have millions of noses is to prevent cells from making a mistake. Because your immune cells are blind and deaf and stupid, they have no way of asking questions. They don't know if a signal is real or if they are interpreting it correctly. For example, a Neutrophil could pick up a cytokine that would be left over from an already-won battle. Being wrong would be a waste of resources or could distract the Neutrophil. The solution is not to rely on one single nose but on many at the same time. Smelling something with one single nose will yield no reaction. A few dozen

---

to function as a sort of guide system for immune cells. So in a nutshell, chemokines are cytokines that are guiding or attracting immune cells to a place. When immunologists talk about "cytokines," they usually mean cytokines that confer information like what is going on in an infection, which type of pathogen has invaded, and which cell is needed to fight it. OK wait, this is getting confusing. Chemokines are cytokines but also cytokines do different things than chemokines? Well, welcome to the world of immunology, where words exist to make your life harder. Here is how we are going to solve this in this book: We will just use the word cytokines because to understand the general principles you need to understand one thing: Cytokines are a diverse group of information proteins that make your immune cells do a lot of different things. One of them is to make them move.

noses smelling something will get an immune cell mildly excited. But a few hundred or even thousands will rile it up pretty intensely and make it react with striking violence!

This principle is extremely important. A signal needs to pass a specific threshold to compel a cell to do something. This is one of the ingenious regulatory mechanisms of your immune system. A small infection with a couple dozen bacteria will only cause a few immune cells to send out a few cytokines and only a few other cells will smell these signals. But if the infection is big and more dangerous, many signals will be sent, and many cells will react. And because there is a lot of battle "perfume" around them, they will react decisively. Not only does the intensity of the smell call more cells to help, it also makes sure that an immune response shuts itself down. The more successful the soldiers at the battlefield are and the fewer enemies are alive, the fewer cytokines the immune cells will release. Over time, fewer and fewer reinforcements will be summoned to the battlefield. And on the battlefield the fighting cells will die over time through suicide. If things go correctly, the immune system shuts itself off.

In some cases this whole system can break down though and have horrible consequences. If there are too many cytokines the immune system can lose all constraint, become super enraged, and overreact massively—which leads to an appropriately named *cytokine storm*. This is nothing more than way too many immune cells releasing way too many cytokines even if there is no danger. But the consequences are horrible. The flood of activating signals wakes up immune cells all over the body, who then might release more. Inflammation rises massively and is no longer limited just to the place of infection. Immune cells flood the affected organs and can cause profound damage. Blood vessels all over the body get leaky and fluid rushes into the tissue and out of the vascular system. In the worst case, the blood pressure will fall to critical levels, and organs will not receive enough oxygen and begin to shut down, which can end deadly. Luckily in your day-to-day you don't need to worry about this too much though; cytokine storms only happen when things go horribly wrong.

We sort of have elegantly skipped over a question, though: How exactly do cytokines convey information and what does this mean? How does a protein tell a cell what to do? As we discussed before, cells are protein robots guided by biochemistry. The chemistry of life causes sequences of interactions between proteins that are called pathways. The activation of pathways causes behavior. In the case of cytokines, the information protein of the immune system, this happens through pathways that involve special structures called **receptors** on cell surfaces. They are the noses of your cells.

In a nutshell, receptors are protein recognition machines that stick in the membranes of cells. A part of them is outside the cell and another part is inside the cell. Actually, about half of the surface of your cells is covered by myriads of different receptors for all sorts of functions, from taking in certain nutrients to communicating with other cells or as triggers for a variety of behaviors. In a simplified way, receptors are sort of the sensory organs of cells that let the insides of the cells know what happens on the outside. So if a receptor recognizes a cytokine, it triggers a pathway inside the cell. A sequence of proteins interacting with each other that end up signaling the genes of the cell to be either more or less active.

In a nutshell, proteins interact with proteins a few times, until eventually, they change the behavior of a cell. The actual biochemistry of the immune system is a nightmare all on its own so we will skip the details here. (Even though it can be pretty cool to learn if you are patient and have a high tolerance for lots of complicated names.)

So to summarize the important parts: Cells have millions of noses on their outsides that are called receptors. They communicate by releasing proteins that carry information, called cytokines. When a cell smells cytokines with their receptors (noses), they trigger pathways inside the cell that change their gene expression and therefore the behavior of the cell. So cells can react to information without being conscious or having the ability to think, guided by the biochemistry of life. This enables them to do pretty smart things even though they are technically very stupid. Some cytokines do also function as a navigation system—an immune cell can smell where they are coming from and literally follow their noses to the battlefield.

Now that we learned how cells sense their environment, there is one last principle that is important for understanding your immune system before we return to the battlefield. How does a cell "know" what a bacterium smells like? Why do bacteria smell like bacteria at all? How does your immune system recognize friend from foe?

11 Smelling the Building Blocks of Life

ONE OF THE FIRST THINGS WE LEARNED WAS THAT YOUR INNATE IMMUNE System distinguishes *self* from *other*. But how does the Innate Immune System know what and who to attack? Who is *self* and who is *other*? And more specifically, how do your soldier cells know what a bacterium smells like? As we discussed earlier, one of the biggest advantages microorganisms have over multicellular animals is the rapid pace at which they are able to change and adapt. As multicellular life has been in competition with microorganisms for hundreds of millions of years now—why didn't bacteria find ways to start hiding their smell? The answer lies in the structures that make living things.

All life on earth is made from the same fundamental molecule types that are arranged in different ways: carbohydrates, lipids, proteins, and nucleic acids. These basic molecules interact and fit together to create structures and these structures are the building blocks of life on earth. We discussed the most important building block already quite a bit, proteins. So for simplicity, we will focus on proteins here as they account for the majority of building blocks—this does not mean the others are not important but the principle is mostly the same and it is helpful to zero in on something.

As we said before, the shape of a protein determines what it can do and how it can interact with other proteins, what structures it can build, and what information it can convey. Every shape is a little bit like a 3D puzzle piece that, together with other pieces, makes up the overall puzzle. Puzzle pieces are a good way to imagine protein shapes because they make something else clear too: Only certain shapes can connect with other certain shapes. But if they do, they fit together quite nicely and firmly. Since there are billions upon billions of different possible protein shapes, life has a large

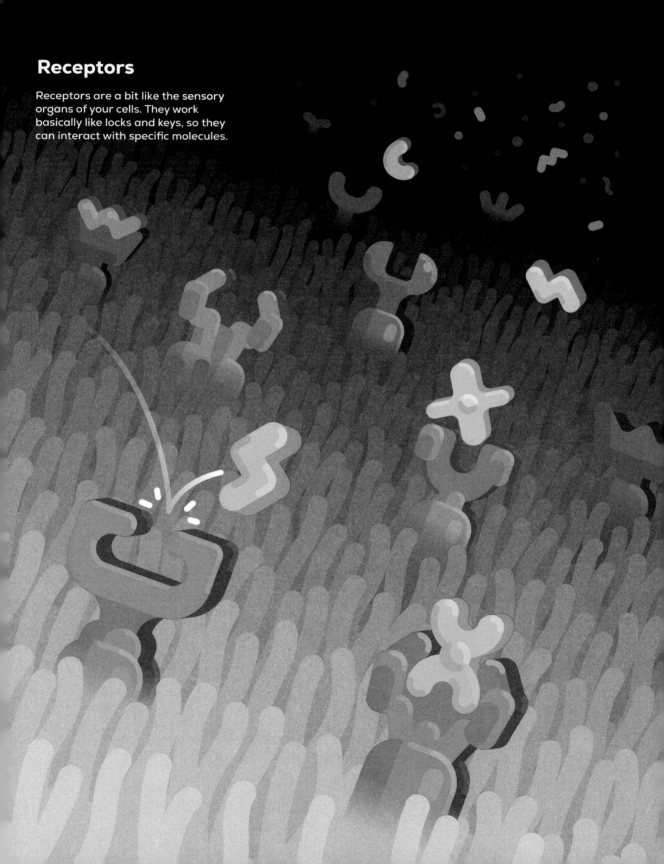

Receptors

Receptors are a bit like the sensory organs of your cells. They work basically like locks and keys, so they can interact with specific molecules.

variety of pieces to choose from when it wants to build a new living thing—say for example, a bacterium. There are a lot of different bacteria you could build from the available protein puzzle pieces of life. Except actually there are limitations to that freedom.

For some specific jobs, the protein puzzle pieces of life can't be altered and still keep their function. It doesn't matter how much a bacterium mutates or what new sort of clever protein combination it comes up with: There are certain proteins that it can't stop using if it wants to be a bacterium. Like, for example, you can make a car in many different forms and colors. But you will not really escape the fact that you need wheels and screws if you want to have a car in the end. It's the same for bacteria. Your immune system is using this fact to recognize if something is *self* or *other*. So how does this work in reality?

A great example is the flagellum. Flagella are micromachines that some species of bacteria and microorganisms use to move. They are long protein propellers attached to the bacteria's tiny butts that are able to rotate fast and propel the tiny being forward. Not all bacteria have them, but many do. It is a pretty ingenious way to move around in the microworld, especially if you are living in shallow and stale water. Human cells do not use them at all.*

---

* OK no, this is actually not true! Sperm cells do use a long and powerful flagellum to move forward (which are technically different structures that work differently but are called the same because hey, biology is not confusing enough apparently). But sperm is a fascinating example in any case. Think about it, why does the body of a woman not recognize sperm cells as other and kill them right away? Well, it does! This is one of the reasons you need about 200 million sperm cells to fertilize a single egg! Right after sperm is delivered into the vagina, it is confronted with a hostile environment that it has to deal with. The vagina is a pretty acidic and deadly place for visitors, so sperm cells move on as fast as possible to escape it. Most of them gain access to the cervix and uterus within a few minutes.

Although here they are greeted by an onslaught of Macrophages and Neutrophils that kill the majority of the friendly visitors that are only trying to do their job. Sperm cells are at least a bit equipped to deal with the hostile immune system (a little like a specialized pathogen if you think about it). They release a number of molecules and substances aimed at suppressing the angry immune cells around them, to buy them a little bit of time. And it may actually be the case that they are able to communicate with the cells that line the uterus, to let them know that they are friendly visitors, which might turn down inflammation. But there is a surprisingly large number of things that are not completely understood yet, in these interactions. In any case, from the millions of sperm cells that entered, only a few hundred enter the fallopian tubes and get a shot at fertilizing the egg.

So if an immune cell recognizes that something has a flagellum, it knows that this something is 100% *other* and has to be killed. So over hundreds of millions of years, the innate immune system of many animals evolved to sort of save the shapes of certain puzzle pieces that are used only by enemies like bacteria. For a lack of a better word, it "knows" that certain puzzle pieces always mean trouble. Of course, your cells don't know anything because they are stupid. But they have receptors! And it so happens that your innate immune cells have receptors that can recognize the protein puzzle shapes that make up flagella, and will enable the immune cells to eliminate them.

The proteins that make up the flagellum of a bacterium are the matching puzzle pieces for the receptors on our immune soldiers. When a Macrophage receptor connects to a bacterium protein that fits, two things happen: The Macrophage gets a tight grip on the bacteria and it also triggers a cascade inside the cell that lets it know that it found an enemy and that it should swallow! This basic mechanism is at the core of how your innate immune system knows who is an enemy or not.

Now, the flagellum protein is not the only kind of protein puzzle piece your immune soldiers can recognize. Your innate immune system can recognize quite a variety of proteins with a few receptors. Just like with cytokines, these special receptors work a bit like sensory organs, like protein recognition machines. It's a very simple mechanism really: Receptors themselves are special puzzle pieces, able to connect to another puzzle piece—which in this case, means the shape of flagella proteins. If the Macrophage is able to connect, it goes into kill mode.

This is how your innate immune cells are able to recognize bacteria, even if they never have encountered a specific species ever before. Every bacteria has some proteins that it can't get rid of. And your innate immune cells come equipped with a very special group of receptors that are able to recognize all the most common puzzle pieces of our enemies:

Toll-Like Receptors—their discovery was worth two Nobel Prizes. "Toll" means "great" or "amazing" in German and it is a very fitting name for this extremely amazing information device. The immune system of all animals has some variant of Toll-Like Receptors, which makes it one of the oldest parts of the immune system that evolved probably more than half a billion years ago. Some toll-like receptors can recognize the shape of flagella, others certain nooks and crannies on viruses, others again telltale signs of danger and chaos, like free-floating DNA.

Neither bacteria, viruses, protozoa, nor fungi can completely hide from these receptors no matter what they do. There are toll-like receptors that do not even have to touch an enemy directly. As we said in the beginning of this chapter, bacteria stink. Just by doing their thing and being alive, microorganisms will sort of sweat out proteins and other garbage that can be picked up by your immune cell receptors, and which give their presence and identity away. As great as it would be for bacteria not to do that, they can't avoid it entirely. Your innate immune system has evolved alongside bacteria for hundreds of millions of years and learned to sniff around for these specific bacteria puzzle pieces. This mechanism enables Neutrophils and Macrophages to detect them, even without knowing what kind of bacteria has entered your body. They just recognize the odor of enemies and that they need to have their heads bashed in.

This principle of cells identifying the puzzle pieces of enemies with sort of sensory receptors on their surfaces is called *microbial pattern recognition* and it will become even more important later on for the adaptive immune system, which uses the same basic mechanism, but in a much more ingenious way.

OK!

Enough explanations of principles. Equipped with this knowledge we can revisit our battlefield and get to know another one of the most powerful and cruel weapons of your innate immune system. A weapon that is tiny even for cells and bacteria.

Remember when you were hiking and stepped on the nail, how that invisible army appeared and just started maiming and killing enemies as fluid from the blood flooded the battlefield during inflammation? Well, it's time to learn what it was. Unfortunately it is cursed with one of the worst names in immunology: The Complement System.

12 The Invisible Killer Army: The Complement System

THE COMPLEMENT SYSTEM IS THE MOST IMPORTANT PART OF YOUR IM-mune system that you have never heard of. Which is pretty weird because so much of your immune system is made to interact with it and if it doesn't work properly the consequences for your health are immense and quite dire.

The complement system is one of the oldest parts of your immune system, as we have evidence that it evolved in the oldest multicellular animals on earth, more than 500 million years ago. In a sense, it is the most basic form of any animal's immune response, but it is also very effective. Evolution does not like to preserve useless things, so the fact that the complement system has been in place for that long, without changing that much, shows how incredibly valuable it is for your survival. Not only has it *not* been replaced as organisms have become more complex, your other immune defenses have become fine-tuned to make it more powerful.

One of the reasons the complement system is largely unknown is that it is mind-numbingly and head-explodingly unintuitive and complex. Even people who need to learn about it in depth in university can struggle with getting a clear mental image of all its different processes and interactions. No part of immunology has been cursed with worse and more difficult-to-remember names for its parts. Luckily, understanding and remembering all its details are wholly unnecessary for anyone who is not studying advanced immunology. So we are going to brush over a lot of details because we can and life is too short for stuff like this. If you are the type of person who would like to know the details, there are illustrated diagrams with all the correct names and mechanisms.

OK cool, but what IS the complement system?

Basically the complement system is an army of over thirty different pro-

Complement Proteins

C3b · **C3a** · **Bb** · **Ba** · **C4b** · **C4a**

C2b · **C2a** · **D** · **P** · **C1q** · **C1r**

C1s · **MBL** · **MASP-1** · **MASP-2** · **C5b** · **C5a**

C6 · **C7** · **C8** · **C9** · **C1INH** · **MCP**

DAF · **H** · **C4bp** · **CD59** · **CR1** · **CR2**

CR3 · **CR4**

One of the key players of our immune system is the complement system. It consists of an army of over 30 different proteins, which work together in a complex and elegant dance to stop intruders. In a nutshell, the complement system does three things: It cripples enemies, activates the immune system, and rips holes in things until they die.

teins (not cells!) that work together in an elegant dance to stop strangers from having a good time inside your body. All in all, about FIFTEEN QUIN-TILLION complement proteins are saturating every fluid of your body right now. Complement proteins are tiny and they are everywhere. Even a virus looks reasonably large next to them. If a cell were the size of a human, a complement protein would barely be the size of a fruit fly egg. Since it is even less able to think and make decisions than your cells are, it is guided by absolutely nothing but chemistry. And yet it is able to fulfill a variety of different objectives.

In a nutshell, the complement system does three things:

- It maims enemies and makes their lives miserable and unfun.
- It activates immune cells and guides them to invaders so they can kill them.
- It rips holes into things until they die.

But how? After all, these are just a lot of mindless proteins, randomly drifting around without will or direction. But this is actually part of the strategy. Complement proteins float around in a sort of passive mode. They do nothing, until they get activated. Imagine complement proteins as millions of matches that are stacked very close together. If a single match catches fire, it will ignite the matches around it, these ignite more, and suddenly you have a huge fire.

In the world of complement proteins, catching fire means changing their shape. As we said before, the shape of a protein determines what it can and can't do, what they can interact with, and in what way. In their passive shape complement proteins do nothing. In their active shape, however, they can change the shape of other complement proteins and activate them.

This simple mechanism can cause self-enforcing cascades. One protein activates another. Those two activate four, which activate eight, which activate sixteen. Very soon, you have thousands of active proteins. As we learned briefly when we talked about the cell, proteins move extremely fast. So within seconds complement proteins can go from being totally useless to an active and unavoidable weapon that is spreading explosively.

Let's look at what this looks like in reality. Think back to the battlefield, 81 the injury from the nail. A huge amount of damage was done and the Macrophages and Neutrophils ordered inflammation, which made the blood vessels release fluid into the battlefield. This fluid carries millions of complement proteins that quickly saturate the wound. Now the first match needs to catch fire.

In reality, this means that a specific and very important complement protein needs to change its shape. It has the amazingly useless name "C3." How C3 changes its shape and activates exactly is complex, boring, and not important right now, so let us just pretend it happens randomly, by pure lucky chance.*†

All you really need to know is that C3 is sort of the most important complement part, the first match that needs to catch fire to start the cascade. When it does, it breaks into two smaller proteins with different shapes that are now activated. The first match is lit!

One of these C3 parts, very creatively called C3b, is like a seeker missile. It has a fraction of a second to find a victim or it will be neutralized and shut itself off. If it does find a target, say a bacterium, it anchors itself very tightly to the bacterium's surface and does not let go. By doing so the C3b protein changes its shape again, which gives it new powers and abilities. (In a way complement proteins are like mini protein Transformers.) In its new form it is able to grab other complement proteins, change their shape, and merge with them. After a few steps it has transformed itself into a recruiting platform.

This platform is an expert at activating more C3 complement proteins that start the whole cycle anew. An amplification loop begins. The cycle of activation and new shapes starts over and over and over. More and more newly activated C3b attaches to the bacteria, creates new recruitment platforms, and activates even more C3. Within a few seconds of the first complement protein activating, thousands of proteins cover the bacterium all over.

* Complement activating randomly, by pure lucky chance, is actually one of the ways it can happen. There are other, more complicated ways complement can activate, but have a look at the fancy diagrams for that!†

† Also, does this random activation happen even when there are no enemies around? Indeed it does! Your cells have defenses against your own complement system to prevent random complement proteins from accidently attacking them!

Alternative Pathway

1. C3 breaks into two proteins C3a and C3b.

2. C3a proteins flood away and alarm Immune Cells.

3. C3b anchors itself tightly to the bacterium.

4. C3b is now able to grab other complement and changes shape.

5. After some steps it has transformed into a C3 recruiting platform.

6. This platform is activating more C3 proteins and the cycle continues.

7. Within a few seconds, thousands of complement proteins cover the bacteria.

C3

C3a

C3b

B

Ba

P

C3 Convertase

Bacterial Cell Wall

For the bacterium this is very bad. Imagine you would go through the day, minding your business, and suddenly hundreds of thousands of flies, in unison, covered your skin, head to toe. This would be a horrifying experience and not something you could just ignore. For a bacterium, this process can cripple and maim it, making it helpless and slowing it down considerably.

But there is more—do you remember the other part that broke off C3? This one is called C3a, because why not, I guess. It is like a distress beacon, just like the cytokines that we discussed two chapters ago. It is a message, an alarm signal. Thousands of C3a flood away from the site of battle, screaming for attention. Passive immune cells like Macrophages or Neutrophils begin smelling them, picking them up with special receptors, and awake from their slumber to follow the protein tracks to the site of infection. The more active alarm complement proteins they encounter, the more aggressive they get, because active complement always means that something bad triggered them. The trail of C3a complement guides them exactly to the place where they are needed the most. In that case, complement does exactly the same job as cytokines, only that they are passively generated and don't need to be generated by cells, like cytokines do.

So far the complement has slowed down the invaders (C3b flies covering their skin) and called for help (C3a distress beacons). Now the complement system begins to actively help to kill the enemy. As we discussed before, the soldier cells are phagocytes, cells that swallow enemies whole. But to swallow an enemy they need to grab it first. Which is not as easy as we made it out to be. Because bacteria prefer not to be grabbed and try to slip away.

And even if they didn't insist on trying to not be killed, there is a sort of physics problem: The membranes of cells and bacteria are negatively charged—and as we learned from playing with magnets, the same charges repel each other. This charge is not so strong that it can't be overcome by a phagocyte, but it does make it considerably harder for immune cells to grab bacteria.

But!

Complement has a positive charge. So when complement proteins have anchored themselves to the bacteria, they act as a sort of superglue, or better, little handles, that make it much easier for your immune cells to grab and hold on to their victims. A bacterium covered in complement is easy prey for soldier immune cells and in a way, much tastier! This process is called *opsonization*, which comes from an old Greek word for a delicious side dish. So if an enemy is opsonized, it is made more delicious.

But it gets even better. Imagine being covered in flies again. Now imagine the flies turning into wasps in the blink of an eye. Another cascade is about to begin. This one will be deadly. On the surface of the bacteria the C3 recruitment platform changes its shape again and starts to activate another group of complement proteins. Together they begin the construction of a larger structure: A *Membrane Attack Complex*, which, I promise, is the only good name within the complement system. Piece by piece, new complement proteins, formed like long spears, anchor themselves deep in the bacteria's surface, impossible to remove. They stretch and squeeze, until they rip a hole into it, one that can't be closed again. A literal wound. Fluids rush into the bacterium and its insides spill out. Which makes it die quite quickly.

But while bacteria are not happy about complement, the enemy it might be the most useful against are actually viruses. Viruses have a problem, namely that they are tiny floaty things and need to travel from cell to cell. Outside of cells, they are basically hoping to randomly bump against the right cell to infect it by pure chance, which also makes them virtually defenseless as they float around. And here complement is able to intercept and cripple them so they become harmless. Without complement, virus infections would be a lot more deadly. But more on viruses later.

Back in our nail-inflicted wound, millions of complement proteins have maimed or killed hundreds of bacteria, making it much easier for your Neutrophils and Macrophages to clean them up. The fewer bacteria the complement proteins find to attach to, the fewer get activated. And so, the complement activity slows down again. When there are no more enemies around, complement just becomes a passive and invisible weapon again. The complement system is a beautiful example of how many mindless things can do smart things together. And how important collaboration is between the different defense layers of your immune system.

OK. In terms of raw brutal fighting power we have gotten to know the most important soldiers of your body and learned about some of the core principles that keep them going and make them work. Let us summarize

Innate Immune System 101:

 Physical Barrier Macrophages Neutrophils Complement

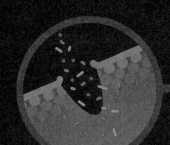

1. Border wall (skin) is breached.

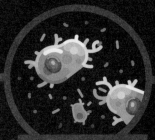

2. Macrophages eat and kill.

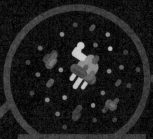

5. Reinforcements including Complement arrive.

4. Immune Cells order Inflammation.

3. Macrophages call Neutrophils.

6. Complement mark, maim, and kill.

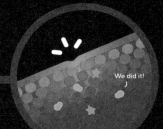

We did it!

7. Invaders are defeated.

briefly what we learned so far about the Innate Immune System before we move on.

Your body is wrapped in an ingenious self-repairing border wall that is incredibly hard to pass and that protects you extremely efficiently. If it is breached, your Innate Immune System reacts immediately. First your black rhinos, Macrophages, huge cells that swallow enemies whole, appear and dish out death. If they sense too many enemies they use cytokines, information proteins to call your chimp-with-machine-gun Neutrophils, the crazy suicide warriors of the immune system. Neutrophils don't live long and their fighting is harmful to the body because they kill civilian cells. Both of these cells cause inflammation, bringing in fluid and reinforcements to an infection, making a battlefield swell up. One of the reinforcements is complement proteins, an army of millions of tiny proteins that passively support the immune cells in their fight and help mark, attach to, maim, and clear enemies. These powerful teams together are enough for most small wounds and infections you encounter.

But what if all of this is not enough? After all, we just assumed all of this would work out. The sad reality is that often enough, it doesn't. Bacteria are not just pushovers but have developed a number of strategies to hide or avoid the first line of defense. Tiny wounds can be a death sentence if an infection is not contained and eliminated.

So let us escalate the situation.

13 Cell Intelligence: The Dendritic Cell

IN THE WOUND CAUSED BY THE RUSTY NAIL, THINGS HAVE STARTED TO GET out of hand. Despite fighting bravely for hours and killing hundreds of thousands of enemies, the Macrophages and Neutrophils could not eliminate the infection. Out of all the different bacteria that invaded the wound, all were maimed, butchered, and eaten, except one species. That one species was not particularly impressed by the defenses and resisted.*

These pathogenic soil bacteria in your infected wound employed defenses, multiplied quickly, and gained a real foothold. They nourish themselves with the resources that were meant for civilian cells and begin defecating everywhere, releasing chemicals that hurt or kill cells, civilians, and defenders. The complement proteins that came with the first waves of fluids from the blood have been largely used up and more and more of the immune cells that fought for hours and days are giving up and dying of exhaustion.

And while fresh Neutrophils still arrive, their reckless fighting becomes more and more of a burden. They order more inflammation, refreshing the complement resistance but also making more and more tissue swell up. The collateral damage is increasing quickly and, at this point, more civilian cells

---

* Hey, you know what might be fun, how about an example of how bacteria resist your immune defenses. Many pathogenic bacteria don't care that much about the complement system, for example. While complement might be super deadly for most bacteria, real pathogens laugh off these silly little proteins and just do their thing in your body, carefully avoiding them. A very fascinating example is the bacterium *Klebsiella pneumoniae*, a pathogen that causes, among other horrible things, pneumonia. It avoids the whole complement affair by hiding itself from complement proteins behind a sticky and gooey structure called a capsule. Which is quite literally a slimy sugary coat the bacteria produce to cover the molecules the immune system would recognize. Simple and effective, like a deodorant for bacteria.

die through the efforts of your immune system than through the actions of the bacteria. On all sides the death count rises rapidly and there is no end in sight.

And now, on the scale of the human, you really begin to notice. You finished the hike minorly annoyed, came home, took a shower, and put a Band-Aid on your wound. But the next day, walking is still a tiny bit unpleasant. Your toe has swollen considerably, it is red, and you feel it throbbing. Even without putting pressure on it, your toe hurts. As you examine it, squeezing it, the crusted wound bursts open and a drop of yellowish pus oozes out.

This strange-smelling substance can emerge from wounds a day or two after they have been infected. Pus is the dead bodies of millions of Neutrophils that fought to the death for you, mixed in with ripped-apart remains of civilian cells, dead enemies, and spent antimicrobial substances. A little bit disgusting, sure, but also a testament to the selfless effort of your immune cells engaged in a fight for your life that has to end in their death. Without your Neutrophils' sacrifice, this infection would have spread already. Maybe to the bloodstream, which would give the intruders access to the whole body and that would be really, really bad.

But there is still hope. While the battle has been raging, the intelligence portion of your innate immune system has been quietly doing its job in the background: The Dendritic Cell is on its way.

For a long time, Dendritic Cells were not really taken seriously, which makes sense if you look at them: they are just ridiculous. Large cells with long starfish-like arms flopping around everywhere, drinking and vomiting constantly. But it turns out they have two of the most crucial jobs of your entire immune system: They identify what kind of enemy is infecting you, if it is a bacterium or a virus or a parasite. And they make the decision to activate the next stage of your defense: Your adaptive immune cells, your heavy, specialized weapons that need to come in if your innate immune system is in danger of being overwhelmed.

Dendritic Cells are very careful and relaxed sentinel cells. They sit around just about everywhere in your body below your skin and mucosa and they are present in all of your immune bases, your lymph nodes. Their job is simply to get drunk. The Dendritic Cell is like a careful connoisseur of the fluids of your body that flows around between your cells. In a way, it treats

Dendritic Cells

A careful connoisseur of the fluids of your body, the Dendritic Cell with its floppy arms constantly swallows and spits out the fluids around it. As soon as its tastes virus or bacteria parts, parts of dying civilian cells, or alarm cytokines it stops spitting and starts swallowing and storing samples. Then it leaves the battlefield and enters the Lymph system to activate the adaptive immune system.

them like expensive wine at an exclusive wine tasting event. It takes a sip, moves it around in its imaginary mouth to get the full picture of all its different tastes and components, and then spits it out again. On an average day, it swallows and spits out multiple times its own volume.

The Dendritic Cell is always looking for a few very particular tastes—the flavor of bacteria or viruses, the taste of dying civilian cells, or the taste of alarm cytokines from fighting immune cells. When it takes a sip and recognizes any of these flavors, it knows that danger is present and goes into a more active sampling mode. Now the Dendritic Cell stops spitting and begins swallowing. It only has a limited amount of time to sample and it is determined to use every second of it. Similar to Macrophages, it begins phagocytosis, grabbing and swallowing whatever garbage or enemies that are floating around the battlefield. But with one major difference—the Dendritic Cell is not trying to digest any enemies. It is still breaking them down into pieces but it is doing so to collect samples and to identify them. Not only is the Dendritic Cell able to distinguish if an enemy is, for example, a bacterium, it can distinguish between different species of bacteria and knows what sort of defense is needed against them.

In your infected toe, for a few hours, this is what the Dendritic Cell did: it floated around a bit, and swallowed as many samples as it was able to grab with its long, weird tentacle arms. It collected and analyzed and stored all sorts of chemicals and enemy corpses it could get. After a few hours its internal timer runs out. Suddenly the Dendritic Cell stops sampling. It has all the information it needs, and since it smells that the battle is very much still active and dire, it begins moving. The Dendritic Cell takes off and leaves the battlefield—its destination is the great gathering place, the intelligence center where millions of potential partners await.

Once a Dendritic Cell is on its way, it has become something like a snapshot of the state of the battlefield, at a particular point in time. A living information carrier of what was going on at the site of infection when they took their samples. We will learn more about this later, but, in a nutshell, the Dendritic Cell delivers *context* to the Adaptive Immune System. If it continued to sample while it was in transit, this could cause two issues: the samples it had collected on the battlefield would be diluted by samples from the journey, and the level of danger would not be as obvious from the snapshot.

And secondly, if the cell sampled outside the battlefield, it could pick up harmless material from your body and accidentally cause an autoimmune disease. You don't need to understand how and why at this point, we will discuss these horrible and fascinating diseases later on.

In any way, the battlefield snapshot, the living information carrier, needs to be delivered to a lymph node. To get there, the Dendritic Cell has to enter the Immune System Superhighway: *The Lymphatic System*, which is a great opportunity to get to know your internal plumbing!

14 Superhighways and Megacities

Consider the continent of the flesh again for a moment, the sheer scale of a human from the perspective of a cell. For a cell, you are a gigantic mountain of meat, more than ten times the height of Mount Everest. Not just a uniform pile of flesh, though, but organized into many different nations and countries that perform the most diverse jobs—from a network of high-voltage electric wires that transfers the orders and instructions of the thinking nation of the brain to the acid ocean of the stomach and the united nations of the gut that process raw resources and transform them into neat food packages, which are then distributed by oceans of fluid filled with delivery swimmers.

And among all of these systems and nations, there is the megacity and superhighway network of your immune system: *The Lymphatic System.* It does not get as much love in textbooks because it is not as clear-cut and obviously useful as the heart with its blood vessels or the brain with its electric wiring. It does not have a giant, central organ like your liver but hundreds of tiny ones. But just like your cardiovascular system, it does have a far-reaching network of vessels and its own special fluid. And without it, you would be equally dead as without a heart. Let's explore it briefly.

Your network of *lymphatic vessels* is miles long and covers your entire body. It is a sort of partner system to your blood vessels and blood. The main job of your blood is to carry resources like oxygen to every cell in the body and to do that, some of the blood needs to actually leave your blood vessels and drain into your tissue and organs to deliver the goods directly to your cells. (Which makes total sense if you think about it for a moment but still feels a bit weird.) Most of that blood is then reabsorbed by your blood vessels. But some of its liquid parts remain in the tissue between your cells and need to be transported back into circulation again. The lymphatic system is responsible for this job. It sort of constantly drains your body and tissues of

The Lymphatic System

Your immune system has its own superhighway system and hundreds of bases.

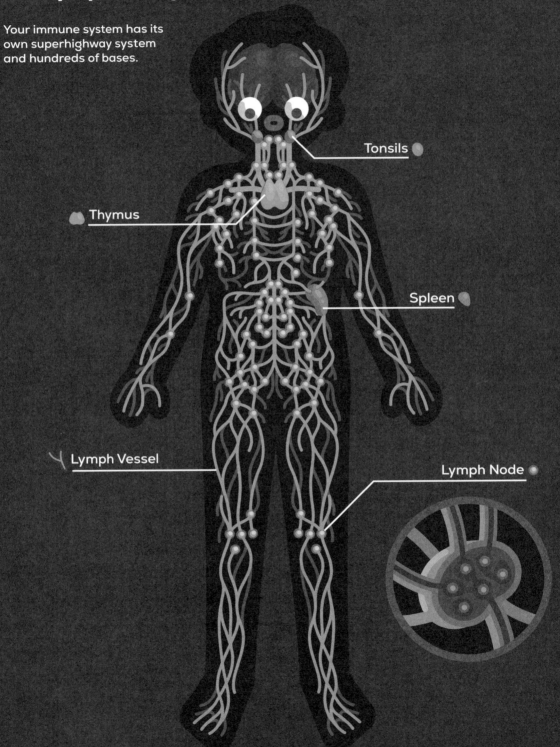

Tonsils

Thymus

Spleen

Lymph Vessel

Lymph Node

excess fluid and delivers it back to the blood where it can circulate again—if it didn't, you would swell up like a balloon over time.

Your lymphatic system begins as a tight and complex network of capillaries spread throughout your tissue. Vessels, bulky and irregular. They are built like a series of one-way valves—water can enter them from the tissue but it cannot flow back. There is only one direction, as very gradually small lymph vessels merge into bigger ones, which then continue to merge into larger ones. Since the lymphatic system has no real heart, the water flows slowly. If a cell were the size of a human, your blood would be like a roaring stream, moving multiple times faster than the speed of sound. In contrast, traveling through the lymphatic vessels would be like taking a chill cruise with sightseeing and no rush.

Your heart pumps and transports nearly 2,000 gallons of blood through your body every single day, while your lymphatic system only transports around three quarts from your tissues back to your blood. This slow movement is made possible by negative pressure and by a very thin muscle layer that surrounds the vessels. You could imagine this as a very thin, spread-out, pump-like pseudo-heart that covers your whole body and only beats once every four to six minutes.*

The fluid transported through the lymphatic system is called *lymph*, and if you find blood a bit disgusting, you won't like lymph either. It is mostly clear, but in some places, like your gut region, it can be yellowish-white and look like old, gross milk. It gets this color because it does not just transport water, it is also your waste management and alarm system. When it drains the excess fluid between your cells, it picks up all sorts of detritus and garbage: damaged and destroyed body cells, dead or sometimes even alive bacteria or other invaders, and all sorts of chemical signals and stuff just hanging out.

This is especially important if you are suffering from an infection because your lymph picks up a sort of cross section of the chemicals that float around battlefields and transports them directly to your immune system

---

* Although "beating" is not a good way to describe it, as the "beats" are not synchronized—it is more like 1,000 toothpaste tubes being squeezed all over your body independently.

intelligence centers, your lymph nodes, where they are filtered and analyzed.*

But while the lymph carries many different things, maybe its most important job is to serve as a superhighway for immune cells. Every second of your life, billions of them are traveling through it, looking for a job. These jobs are given out in the megacities of your immune system that the lymph must pass through before it can become part of the blood again: Your bean-shaped *Lymph Node* megacities—the organs of your immune system. You have around 600 of them spread all over your body.

Most surround your intestines, are in your armpit, neck, and head region, or are near the groin. You can try to touch them right now. Put your head back and carefully feel for them in the soft region below the corners of your jaw. If they are too small to feel right now, you can feel them for sure when you have a sore throat or a cold as they will swell and feel like weird, firm lumps. The lymph node megacities are like huge dating platforms where the Adaptive Immune System meets the Innate Immune System for hot dates. Or better, adaptive immune cells go and look for their ideal match. This is where the traveling Dendritic Cell arrives from the battlefield after about a day of chill travel.

An Aside The Spleen and the Tonsils— Super Lymph Node Best Friends

ONE PART OF YOUR LYMPHATIC INFRASTRUCTURE IS A SPECIAL LITTLE ORGAN that most people are not really aware of although it is pretty important. *The Spleen* is a sort of large lymph node, about the size of a peach but bean shaped. Just like lymph nodes it is a sort of filter but with a much larger scope. For one, the spleen is the place in your body where 90% of your old blood cells are filtered and recycled when their life comes to an end. On top

---

* And funny enough, and too weird not to mention, the Lymphatic System is your fat transportation system. It picks up food fats around your intestines and dumps them into the bloodstream to be further distributed.

of that your spleen stores an emergency reserve of blood, about a cup, which is invaluable if something bad happens and you could use a bit of extra blood in your body. This is still not all, 25 to 30% of your red blood cells and 25% of your platelets (remember, the cell fragments that can close wounds) are stored here for emergencies.

But the spleen is not just an emergency blood storage for injuries but also one of the centers of your soldier cells, a sort of barracks. The main home for another immune cell we did not mention before even though it did help out during the cut: the Monocyte. Monocytes are basically reinforcement cells that can transform into Macrophages and Dendritic Cells. About half of them patrol in your blood right now where they represent the largest single cell that floats through your cardiovascular system. If you suffer an injury and an infection that drains and kills a lot of your Macrophages, they come in as backup. Once they enter the site of infection, they stop being Monocytes and transform into fresh Macrophages. This way, even if you lose a lot of Macrophages to an intense battle, you have a fresh influx of them that does not run out.

The other half of your Monocytes sit in your spleen as a reserve emergency force. While it's easy to think of Monocytes just as replacement Macrophages, there are Monocyte subclasses that have more specialized jobs like working as inflammation superchargers, or they are called to the heart during a heart attack to help the heart tissue with healing itself.

On top of serving as an emergency reservoir and a barracks, the spleen really is just a huge lymph node that filters your blood (and not your lymph fluid, like your regular lymph nodes do) and does all the things lymph nodes do. So when we discuss the function of lymph nodes in more detail, just remember that your spleen does the same thing, only with your blood!

People regularly lose their spleens, for example after traffic accidents where a strong blow to the torso can rupture the small organ critically, so that it has to be removed. Surprisingly this is not as deadly as you might think. Other organs like the liver, regular lymph nodes, and your bone marrow can take over most of its jobs. And about 30% of people happen to have a second spleen that is tiny but will grow and take over the job if the first one is removed.

But it is like, not great to lose your spleen because as you may have thought, most organs in your body exist for a reason. Patients who lose their spleen often become vastly more susceptible to certain diseases like pneumonia, which can be deadly in the worst case. So while it is no death sentence to lose this tiny, weird organ, try to keep it if you can!

The Tonsils are known to people only as weird lumpy things in the back ends of their throats that sometimes have to be surgically removed in children. But they are not just annoying bits of useless tissue. Your tonsils are something like a center of your immune system intelligentsia in your mouth. A lot of different immune cells that we will get to know in this book work here to keep you healthy. To get samples to them, your tonsils have deep valleys where tiny pieces of food can get stuck. Microfold Cells, very curious cells that grab all sorts of stuff from your mouth and pull it deep into the tissue, where they show it to the rest of the immune cell to check out.

This is useful in two ways basically: At a young age this trains your immune system so it can recognize what kinds of the food you eat are harmless and should not be reacted against. And to be able to produce weapons against invaders if it finds any. We will explain all these mechanisms in detail in the rest of the book, so let us not venture in too deep here. If your tonsils are overeager and work too hard, they can become chronically inflamed and swell up, which can cause all sorts of unpleasant symptoms. Sometimes this makes it necessary to remove them surgically, but this depends on the case and is generally not a huge deal if the patient is older than seven years old and has a solid immune system. In a nutshell, what you really need to know about your tonsils is that they are immune bases that actively sample what comes into your body.*

OK, time to get back to our battlefield! Let us be mysterious for a moment. The Adaptive Immune System awakens. Very slowly, like a teenager who was awoken by their mother before sunrise, it stretches and groans as it slowly slides out of bed and gathers its strength.

Back at the site of infection, its presence is desperately needed.

---

* Before the tonsils were understood better it was standard procedure and quite common to remove them if they got infected, or sometimes even as a precaution. Nowadays removing them is considered much more carefully as they do serve a purpose. Pretty amazing if you think about it, how easily humans were on board with removing living parts because they seemed to be annoying and not that useful.

15 The Arrival of the Superweapons

BACK AT THE RUSTY NAIL BATTLEFIELD, THE FIRST DENDRITIC MESSENGERS, armed with snapshots and information, left days ago, an eternity in cell time. The soldiers of the Innate Immune System have been fighting the pathogenic soil bacteria that invaded your tissue vigorously, all this time. By now they must have killed millions of them. Pushing them back over and over, only for the bacteria to spread into more surrounding tissue and resurge with fresh forces. The battlefield is a chaos of dead civilian and soldier cells, NETS erected by Neutrophils (you know, these suicide traps that literally look like nets), toxins and feces of bacteria, alarm signals, and spent complement proteins. Death is everywhere. Millions of immune cells have fought to their demise. All in all, the Innate Immune System will probably win this battle eventually. But it might take weeks and victory is far from certain as there is still the possibility that the immune system will lose and that the invaders will make their way deeper into the flesh giant, causing more mayhem and destruction.

Exhausted by a seemingly endless war, a spent Macrophage slowly moves over the battlefield looking for bacteria to kill. But it is almost done. The Macrophage is so, so tired. All it wants to do is to stop fighting and give up, embrace the sweet kiss of death, and go to sleep forever. It is about to do so, but then it notices something. Thousands of new cells arrive at the battlefield and spread out quickly. But these are not soldiers.

These are Helper T Cells!

Specialist cells from the Adaptive Immune System were forged just for this particular battle and they exist only to fight this specific soil bacterium that has been giving the soldiers so much trouble! One of these Helper T Cells moves around a bit, sniffing and taking in the environment. It seems to collect itself for a moment. And then it moves directly towards the tired Macrophage and whispers something, using special cytokines to convey its message. Suddenly a jolt of energy shoots through the Macrophage's bloated body. In a heartbeat, its spirit comes back and it feels fresh again. But there is something else: A hot, white anger. The Macrophage knows what it needs to do: Kill bacteria, right now! Invigorated, it throws itself against the enemies to rip them into pieces. All over the battlefield this begins to happen as Helper T Cells whisper magic words to tired soldiers, motivating them to get themselves together and engage the bacteria again, with even more violence than before.

But this is not all that is happening. Something weird is going on. Another tiny army—this time directly made by the Adaptive Immune System—has joined the fight. Counting in the millions, it floods the battlefield, dashing against the enemies. The specialist forces of the Antibodies have arrived! Although they are made from proteins just like complement, antibodies are very different.

If complement fights like warriors with clubs and claws, antibodies fight like assassins with sniper rifles. In this case, their purpose is to maim and disarm this exact kind of bacteria that is present right now at the site of infection. There is no escape this time. Bacteria hiding behind cells or trying to escape begin to jerk around as they are being swamped by thousands of antibodies attaching to them. Even worse, multiple bacteria are glued together, unable to move or flee.

With help from the Antibodies, your soldiers can suddenly see them much more clearly and they now seem much more tasty than before, now that they have been opsonized.

Even the complement system now seems to be more aggressive than before, as it once again begins attacking and ripping holes into victims. What had been a desperate and brutal battle for days now turns into a one-sided slaughter very quickly. The pathogenic bacteria have nothing to counter the coordinated tactic of the immune system. Step by step they are eradicated and exterminated without mercy.

At some point the last panicked bacterium is devoured whole by the once-tired Macrophage. The battle is won. Now the cytokine whisper of the T Cells slowly subsides and the Macrophages start to feel tired. The soldiers

The Super Weapons Arrive

When the adaptive immune system
arrives at the battlefield the invaders get
cleaned up fairly quickly.

Neutrophil NET

Enraged
Macrophage

Opsonized and
Clumped up Pathogens

Helper T Cell

Antibodies

around it, mostly Neutrophils that fought so bravely, start killing themselves. Their presence is no longer needed and they know that they would do more harm than good moving forward. The remains of their bodies are cleaned up by fresh young Macrophages that will take their place as the new guardians of the tissue.

Their first job is to help the civilian cells to heal the wound by sending encouraging messages that motivate them to rebuild. Most of the Helper T Cells join the controlled mass suicide but some remain at the former site of infection and settle in to protect the tissue from a future attack.

The inflammation retracts and blood vessels constrict again, while the excess fluid leaves the now-former battlefield, transported away through the lymphatic vessels. The bloated tissue constricts slowly to its former dimensions. The damaged tissue is already regrowing, young civilian cells take the place of the fallen. Regeneration is on its way.

On the human scale, a few days after your unfortunate encounter with the rusty nail, you wake up and notice that your toe is much better. The swelling has gone away, the wound has grown over and left nothing but a faint red mark. Business as usual. Wounds heal, no big deal. You were completely unaware of the drama your cells had to deal with. For you, the whole ordeal was a slight annoyance, while for millions of your cells it was a desperate fight of life and death. They did their duty and gave their lives to protect you.

What happened here? How were the reinforcements from the Adaptive Immune System able to swing the situation at the battlefield so massively and decisively that the bacteria were wiped out? And while you certainly don't want to complain, how come your immune system took its sweet time to get there?

16 The Largest Library in the Universe

It was no coincidence that when the Adaptive Immune System showed up, the desperate battle turned into a brutal bloodbath that devastated the invading bacteria. They never had a chance because the reinforcing cells and Antibodies were born to fight them specifically. Right now your Adaptive Immune System has a specific weapon against every possible enemy in the universe. For every single infection that has ever existed in the past, for all of them in the world right now, and for every single one that might emerge in the future but does not even exist yet. In a way, the largest library in the universe.

Wait. What? How? And why? Well, because it is necessary.

Microorganisms have a huge advantage over us meat giants. Consider how much effort it takes you to make even a single copy of yourself and your trillions of cells. To multiply you first need to find another flesh giant that thinks you are cute. Then you need to go through a complicated dance that hopefully leads to the merger of two cells from you two.

And then you need to wait for months and months while the merged cell multiplies over and over and over, until it has become a few trillion cells and is released into the world as a hopefully healthy human. And even then you only have produced a single mini human that is actually pretty weak and needs years of attention and care before it stops being totally useless. It takes even more years before the offspring can repeat the dance and multiply again. Any sort of evolutionary adaptation to a new problem is super slow with our very inefficient ways.

A bacterium consists of one cell. And it can produce another fully grown-up bacterium in about half an hour. Not only does this mean that bacteria can multiply orders of magnitude faster, they also change orders of magnitudes quicker than you can. For a bacterium you are not a person but a hostile ecosystem applying selective pressure. Your immune system can exterminate thousands and millions of them, but by pure random chance from time to time, there will be an individual that adapts to your defenses and becomes a pathogen: A microorganism that causes disease, as we saw in our battle. Worse still, even in the midst of an ongoing infection, the genetic code of invaders can change in ways that make them harder to kill. Bacteria are a lot of things but they are not weak—the most dangerous ones have evolved ingenious ways to avoid our defenses over the years, and, given the chance, they will improve them even more. So against the powerful enemies from the world of microorganisms, you, as a huge mountain of cells simply can't rely on your innate defenses alone.

And so, to survive these ever-changing enemies that exist in hundreds of millions of varieties, you need something that can *adapt*. Something *specific*. A weapon for every single different enemy. And weirdly enough, your immune system has exactly that. But this seems impossible. How can your slow continent of flesh adapt to create specific defenses for each of the millions of different microorganisms and the millions more that don't even exist yet?

The answer is as simple as it is baffling: The immune system does not so much adapt to new invaders as it already was adapted when you were born. It comes preinstalled with hundreds of millions of different immune cells—a few for every possible threat that you could possibly encounter in this universe. Right now you have at least one cell inside you that is a specific weapon against the Black Death, any variant of the flu, the coronavirus, and the first pathogenic bacteria that will emerge in a city on Mars in one hundred years. You are ready for every possible microorganism in this universe.

What you will now learn may be the most amazing aspect of your immune system. It will take us a few chapters and will introduce not only mind-blowing principles that keep you alive, but also your best defense cells and things like Antibodies, something we hear about quite regularly in the media, especially in the wake of the novel coronavirus.

17 Cooking Tasty Receptor Recipes

To understand how your adaptive immune cells are able to recognize every possible enemy in the universe, let us go back to one of our earlier chapters: "Smelling the Building Blocks of Life." Let's freshen up our memories a bit because the principles we encountered there are crucial to understanding the next part.

As we discussed before, all living things on earth are made from the same basic parts, but mostly proteins. Proteins can have countless different shapes, which you can imagine as 3D puzzle pieces. To recognize a bacteria and to grab it, your immune cells need to connect to the protein puzzle pieces of bacteria.

Your Innate Immune System is able to recognize some of the common protein puzzle pieces our enemies use with those special receptors we discussed called *Toll-Like Receptors*. But this somewhat limits the range of the Innate Immune System, as it is only able to recognize the structures that can connect to the Toll-Like Receptors. Nothing more, nothing less.

While microorganisms can't totally avoid using some of these common proteins, they still have a huge set of other available proteins to use as construction material. In the language of immunology, a protein piece that is recognized by the immune system is called an *antigen*. There are hundreds of millions of possible *antigens* that your Innate Immune System doesn't recognize and through the magic of evolution there will always be new ones created in the future. Antigen is one of these important ideas that will be very relevant for the rest of the book, so one last time so you remember more easily: *An antigen is a piece of an enemy that your immune system can recognize.*

There are hundreds of millions of potential antigens, hundreds of millions of different possible proteins. To solve this problem your adaptive immune system has an ingenious solution. In your body right now, there is at least one immune cell that has a *receptor* that can recognize one of the many millions of different antigens that can exist in the universe. Let us repeat that: *For every possible antigen that is possible in the universe, you have the potential to recognize it inside you right now.*

Let that sink in for a moment. It is sort of easy to brush over this fact without giving it the appropriate amount of wonder. What a weird tactic and even weirder that it works.

But wait a second. Receptors are made from proteins and as we discussed earlier, a gene is the code to build one protein. If you have hundreds of millions of different receptors for every possible protein shape in the universe, do you have hundreds of hundreds of millions of genes only for immune cell receptors? Well, no. The human genome only has about 20,000 to 25,000 genes. Wait. How can you get such an enormous variety of receptors if our genetic code is so much smaller? And it gets even better: Most of your 20,000–25,000 protein-encoding genes are doing other things that are not related to the immune system—like making the proteins that keep the cell alive. In order to generate the largest known library in the universe, evolution has given your immune system a small number of gene fragments, not even whole genes, but fragments. How is this possible? The answer is a deliberate mix and match of those fragments to create a stunning diversity. Let us try to understand how this is possible.

Imagine you were a cook for the most fantastical dinner party in the universe. There are a few hundred million possible dinner guests. And these guests are extremely picky and annoying. Each of them requires a unique and specific recipe for their dinner. If they don't get it, they get annoyed and try to kill you. And to make it harder, you don't know which guests will visit the dinner party beforehand. So you need to get creative.

You look through your kitchen and find only eighty-three different ingredients in total, split into three categories: vegetables, meats, and carbs. If you are confused right now, ingredients represent gene segments! But anyway, you decide to start mixing ingredients to make different recipes.

So for starters, you have fifty different vegetables: Tomatoes, zucchini, onions, peppers, carrots, eggplant, broccoli, and so on. You choose one. Then you move on to meat, which is pretty simple because there are only six options: Beef, pork, chicken, lamb, tuna, or crab. You get one of those. And last you choose one carb out of twenty-seven different possibilities: Rice, spaghetti, fries, bread, baked potatoes, etc. With three different categories and a number of options each, you get recipes like this for example:

Tomato, Chicken, Rice
Tomato, Chicken, Fries
Tomato, Chicken, Bread
Zucchini, Beef, Spaghetti
Zucchini, Chicken, Spaghetti
Zucchini, Lamb, Penne
Onion, Pork, Baked Potato
Onion, Tuna, Fries
Onion, Pork, Fries
Et cetera

And so on. You get the idea. All in all, from only eighty-three different ingredients, by combining them in all possible variations, you get 8,262 unique recipes for your main course! A lot, but not enough to have a unique one for every potential guest who could show up!

So you decide to add a dessert course. You do the same thing again, with fewer ingredients this time but the same principle applies:

Chocolate, Cinnamon, Cherries
Caramel, Cinnamon, Cherries
Marshmallow, Nutmeg, Strawberries
Et cetera

And so on until you get another 433 desserts by combining different sweets and spices! You can randomly combine them with your main courses to get even more variety. So multiplying 8,262 entrees with 433 desserts, you get 3,577,446 unique dinner combinations for your guests! Now that you have millions of dishes, you decide to run wild and use them as the basis for your dinner party. You randomly add or remove parts of the ingredients. For example, for some recipes you cut away half an onion while for others you add a tomato. Every single possible action lets the number of potentially different dishes explode. One of your final recipes could look like this:

Tomato, chicken, rice, half an onion as entree, marshmallow, pepper, strawberries, and a quarter banana as a dessert.

After a long day of cooking and randomly combining or subtracting ingredients, you get at the very least *billions* of unique meals, which is enough for those 100 million possible dinner guests. Most of them taste weird. But the goal was variety for your complicated guests and not taste.

In principle, this is what your adaptive immune cells do with gene fragments. It takes gene segments and randomly combines them, then it does the same again, and then it randomly pulls out or ads in parts, to create billions of different receptors. They have three different groups of gene fragments. They randomly choose one from each group and put them together. This is the main course. Then they do this again, but with fewer fragments for the dessert. And then when they are done, they randomly remove or add in parts. This way your adaptive immune cells create at least hundreds of millions of *unique receptors.*

Each of them fitting one possible guest at the dinner party, which in this case is an *antigen from a microorganism* that could invade your body. So through controlled recombination, your immune system is prepared for every possible antigen an enemy could make. But there is a catch—this ingenious way to create such stunning variety makes your adaptive immune cells critically dangerous to you. Because what is stopping them from developing receptors that are able to recognize *self*, the parts of your own body? Well, their education is.

So let us finally talk about your most important organ you have never heard of.

18 The Murder University of the Thymus

GOING TO SCHOOL OR COLLEGE CAN BE QUITE UNPLEASANT AND ANNOYING. There are schedules, tests, and pressure to perform, other people, and getting up early. And all this while you are transforming from a teenager, the worst stage in the human life cycle, into, ideally, a functional human being.

But human school is harmless, laughable even, compared to the university your Adaptive Immune Cells have to graduate from: *the Murder University of the Thymus.* Your Thymus is absolutely crucial for your survival and, in a way, will decide at what age you will die, so you might think that it would be as well-known as the liver, lungs, or heart. But weirdly enough, most people are not even aware that they have this organ. Maybe because it is pretty ugly.

The Thymus is an unappealing and boring collection of tissue that looks a little bit like two old, lumpy chicken breasts sewn together in the middle. Despite its ugliness, it is one of your most important immune cell universities (others include your bone marrow for your B Cells, for example, but we'll ignore them here as they get their own chapter later). Some of your most powerful, crucial adaptive immune cells are educated and trained here: **T Cells.**[*]

We met one group of T Cells on the battlefield briefly, when they came rushing in to swing the battle, although we haven't even begun to discover all their qualities. T Cells do a variety of things, from orchestrating other im-

---

[*] Actually T Cells got their name from the Thymus, because they go to school here! It's a weird naming convention if you think about it. Imagine you would be called "NW Human" and your sister "B Human" because you went to Northwestern University and your sister went to Brown University.

The Thymus

The Thymus is the Murder University that every T Cell has to pass. Not just to make their parents proud, but to stay alive.

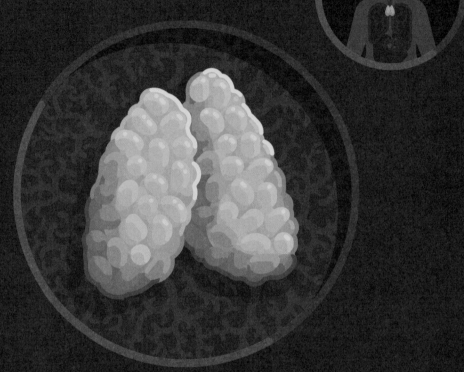

T-Cell Training

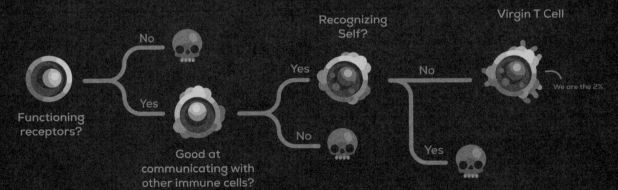

Recognizing Self?

Virgin T Cell

No

We are the 2%.

Yes

Functioning receptors?

No

Yes

Good at communicating with other immune cells?

Yes

No

mune cells, to being antivirus superweapons, to killing cancer cells. We will talk about this amazing cell and all the mind-blowing things it does in more detail later, for now just remember: Without T Cells you are quite dead—they may be the most important Adaptive Immune Cell you have. But before they can fight for you, they need to pass the horribly dangerous curriculum in the Thymus. Failing a test here doesn't mean bad grades. Failing here means death.

Only the best of the best students avoid this fate. As we discussed in the last chapter: *The adaptive immune system is mixing gene segments to produce an amazing variety of different receptors, able to connect to every possible protein, in this context called an antigen, in the universe. This means that each individual T Cell is born with ONE specific type of receptor, able to recognize ONE specific antigen.* But there is a vital flaw: With so many different receptors there are guaranteed to be a large number of T Cells with receptors that are able to connect to proteins from your own cells. This is not a theoretical danger, but the cause of a number of very real and serious diseases that millions of people are suffering from right now called autoimmune diseases.

For example, let's say a *T Cell receptor* can connect to a protein on the surface of a skin cell, it would not understand that it is connecting to a friend. It would just try to kill it. Or worse, since there are quite a lot of skin cells in the human body, it would think a large attack was going on with enemies everywhere, and alert the rest of the immune system to go into attack mode, and cause inflammation and all sorts of chaos. While this is bad enough, it could also affect heart cells or nerve cells, leading to even more dangerous conditions.

At least 7% of all Americans suffer from autoimmune diseases but we will learn more about them later. In a nutshell, an autoimmune disease is your adaptive immune system thinking that your own cells are enemies, that they are *other*. It is no hyperbole to say that this is a critical danger to your survival.

As you can imagine, the body takes this issue extremely seriously and came up with the Murder University of the Thymus to address it. After a fresh and young T Cell has been born it travels to the university and begins its training, which consists of three steps, or better, three tests:

The first test is basically just making sure the T Cells have the ability to make working T Cell receptors. If this were a regular school, this would be the teachers checking if the students have all their notebooks and required reading material with them—only they would not send the students home if they forgot something, but rather, shoot them in the face.*

The T Cells that pass test one have functional receptors. Great job so far! The second test is called *positive selection*: Here the teacher cells check if the T Cells are really good at recognizing the receptors of the cells they will need to work with. Imagine this part as if the teacher is checking if the pens the students brought are all full of ink and that the workbooks are in fine condition. Once again, death is the punishment for failing the second test.

After the two first hurdles have been taken the last and most important test waits for our T Cell students: *Negative selection*. And this might be the hardest test of all. The final exam is simply: Can the T Cell recognize *self*? Can its receptor connect to the main proteins inside the body? The proteins that make you, you? The only acceptable answer is "No, not at all."

So in the final exam the T Cells are presented with all sorts of protein combinations that are used by the cells of your body. The way this happens is pretty fascinating by the way—the teacher cells in the Thymus that do the testing have a special license to make all sorts of special proteins that usually are made only in organs like the heart, pancreas, or the liver and also hormones, like insulin for example. This way they can show the T Cell all kinds of proteins that are marked as "*self*." If a T Cell is able to recognize any of these self-proteins, they are taken out back and shot in the head immediately.†

All in all, 98 of 100 students that enter the university will not survive the training and are killed before they graduate. Roughly ten to twenty million T cells will leave your Thymus today. They represent the successful 2% of

---

* OK, technically no T Cell is killed in the Thymus. To be more correct would actually be that they are told to kill themselves by the teacher cells. So they are ordered to commit suicide. But hey, semantics.

† There is one small exception that could save some of the worst students, that we will get to know later, but in a nutshell, a T Cell that is somewhat good at recognizing "self" can be turned into a special cell called a Regulatory T Cell that has the purpose of calming the immune system down and prevent autoimmunity. But more on this cell later.

survivors. These survivors are so diverse that you end up with at least one T Cell that can recognize basically every possible enemy that the universe could throw at you.*

Unfortunately your Murder University is already in the process of shutting down. Your Thymus basically begins shrinking and withering away when you are a small child. A process that is sped up once you reach puberty. Every year you are alive more and more Thymus cells turn into fat cells or just worthless tissue. The university closes more and more departments and gets worse as you age, until around the ripe age of eighty-five, your T Cell university closes its gates for good. Which is sort of horrible if you like the concept of being alive and healthy. There are other places in the body where T Cells can be educated, but for the most part from this point forward your immune system is more limited than before. Because once your Thymus is gone, you have to get by with the T Cells you have trained up to this point. The absence of the immune cell university is one of the most important reasons why seniors are much weaker and more susceptible to infectious diseases and cancer than younger people. Why is that so? Well, the problem is nature does not care about us once we are no longer making babies, so there is just no real evolutionary pressure to keep us around in older age.†

OK. So in the last two chapters we learned that our adaptive immune system has the largest library in the universe. We learned that after they are born, your T Cells rearrange a few select gene fragments to create billions of different receptors (each T Cell carries only one receptor type). And that in total these many different T Cells, each with its own unique receptor, are able to recognize every possible antigen in the universe. To make sure your

---

* Are you wondering what is happening to all the students that die? There are a lot of Macrophages in your Thymus and their job is to eat all the unlucky fellows who did not pass the test.

† Some of the more promising efforts of the life-extension community are in finding ways to delay the shrinkage or even regrow Thymus tissue. As of the writing of this book one successful study has been conducted with volunteers that claims to have successfully regenerated Thymus tissue—although it had only a very small sample size and its results have yet to be repeated and confirmed by more studies and with more participants. But if you are reasonably young while reading this, there might be a chance that by the age you retire there will be drugs or treatments to regenerate your Thymus!

own adaptive immune cells do not accidentally recognize and attack your own body, these T Cells have to undergo a rigorous training that only a tiny minority survives. But in the end you get a few immune cells for every possible enemy that could ever infect you.

OK, all of this sounds great. But, as with everything in life, there are, of course, a few more tiny problems.

19 Presenting Information on a Gold Platter: Antigen Presentation

As we witnessed in the simple toe infection, having only a few immune cells is not very useful in a full-on invasion. You need hundreds of thousands, if not millions of immune cells to fight a strong enemy effectively. And while your adaptive immune system has billions of different cells, each with a receptor for every possible enemy, it only has maybe ten to a dozen cells with each unique receptor.

Which makes sense if you think about it. If you had millions of cells for every single one of the hundreds of millions of possible different pathogens, you would consist of quadrillions of immune cells and nothing else. On the one hand you would probably never get sick because you were so well prepared. But then again you would just be a puddle of slime. Surviving alone is boring, so nature found a much better and extremely elegant way to solve this conundrum.

When an infection occurs, your immune system determines which specific defense is needed and how much of it. The adaptive immune system works together with the Innate Immune System to find the few cells that have the right receptors for this specific invasion, locate them among billions of others in your huge body, and then rapidly produce more of these cells.

Not only does this method enable you to get by with just a few cells for each possible enemy, it also makes sure that the immune system does not overproduce weapons and waste resources—which is good because the immune system is already a pretty energy-intensive system as is. How does it do this? By preparing a *presentation*.

Your Adaptive Immune System doesn't make any real decisions about who to fight and when it is time to activate—this is the Innate Immune System's job, and this is where the Dendritic Cell, this big, weird-looking cell with many octopus-like arms that takes samples, comes into play. When an infection happens it covers itself with a selection of the enemy's antigens and tries to find a Helper T Cell that is able to recognize one of the antigens with its specific receptors. And this is exactly the reason why the Dendritic Cell is so crucially important. Without Dendritic Cells, there would be no second line of defense. The toe infection battle scene would have had no late-stage turnaround.*

For the first few hours of an infection, the Dendritic Cell samples the battlefield and collects information about the enemy, which is a nice way of saying that it swallows enemies and disassembles them into their parts, or *antigens*. The Dendritic Cell is an *antigen-presenting cell,* which is a complicated way of saying: "covering yourself in your enemies' guts." Dendritic Cells literally disassemble pathogens into antigen-sized pieces and pack them into special contraptions on their membranes. On the human scale this would mean killing an enemy soldier and then covering yourself with bits and pieces of their muscles, organs, and bones so others could examine them. Extremely brutal, but for our cells, pretty efficient and a normal weekday.

Covered in guts, the Dendritic Cell then travels through the lymphatic system to *present them to the Adaptive Immune System, or more precisely, to Helper T Cells.*

All antigen-presenting cells have one thing in common: A very special molecule that is as important as the Toll-Like Receptors and thus deserves to be talked about, even though it has one of those worst names in immunology: *Major Histocompatibility Complex class II.* Or in short, *MHC class II,* which is a bit better but not much.

You can imagine the MHC class II receptor as a *hot dog bun* that can be filled with a tasty *wiener.* The wiener in this metaphor is the antigen. The

---

* Let us use this moment to drive another thing home: Cells are stupid. Dendritic Cells are stupid. Nobody is making any kind of decision or conscious analysis here. The things we are describing here are happening by chance. The magic of your immune system is that it has evolved a setup that increases the chances of these seemingly impossible events to a degree that they are an actual, proper protection! We will explore how this works in more detail in the following chapters.

Antigen Presentation or "Hot Dogs"

1. A bacterium is captured and engulfed via Phagocytosis.

2. The bacterium is ripped into small pieces, called antigens. (The wiener in our hot dog story)

3. The antigen is now loaded on MHC class II molecules. (The bun in our hot dog story)

4. The MHC II molecule now travels to the surface to present the antigen to a helper T-cell.

1.

2.

Antigen

3.

MHC class II

4.

MHC class II molecule:
The Hot Dog Bun

Antigen:
The Wiener

Dendritic Cell

MHC hot dog bun molecule is so important because it represents another security mechanism. Another layer of control.

As we mentioned briefly before and will talk about in detail in the next few chapters, the cells of the Adaptive Immune System are extremely powerful. Activating them by accident has to be avoided at all costs—so a few special requirements have to be met before they are activated. And one of them is the MHC class II receptor, the hot dog bun.

Helper T Cells are able to recognize an antigen only if it is presented in an MHC class II molecule. Or in other words, they eat a wiener only in a hot dog bun. Think of Helper T Cells as really picky eaters—they would NEVER even think to touch and eat a wiener that floats around by itself. No sir, that would be disgusting! Helper T Cells only ever consider eating a wiener if it is nicely presented to them in a hot dog bun.

This makes sure that Helper T Cells can't just get activated by accident because they pick up antigens that float around freely in the blood or lymph. They need to be presented with an antigen that sits in an MHC class II molecule, from an antigen-presenting cell. Only this way does the Helper T Cell have confirmation that there is actual danger and that it should get active!

OK, this is pretty weird stuff and it is OK if it still feels counterintuitive to you. Let's do this again but this time follow a Dendritic Cell from our story with the rusty nail to see how this process works.

So back at our battlefield where soldiers are engaged in an epic battle, Dendritic Cells swallow a cross section of everything floating around, including enemies. If they grab a bacterium they rip it into small pieces, antigens (the wieners), and put them into MHC class II molecules (the hot dog buns) that cover its outsides. The cell is now covered in tiny parts of dead enemies and detritus from the site of infection.

Then the Dendritic Cell makes its way through the lymphatic system to the closest lymph node to look for a Helper T Cell. Remember how in the Lymph Node megacities there are those special dating areas? Places for Dendritic Cells from battlefields and Helper T Cells that travel around the body to meet and find love? Well, let's drop in on such a date.

Our Dendritic Cell, covered in antigens (wieners) that sit in MHC class II molecules (hot dog buns), goes from T Cell to T Cell and rubs its antigen-covered body against them, to see if this yields any reaction. When a Helper

T Cell happens to have the right T Cell receptor, with just the right shape that recognizes the antigen in the MHC class II molecule, it will connect to it. Just like two puzzle pieces snapping perfectly together with a loud click.

This is a pretty exciting moment. The Dendritic Cell actually has found the right Helper T Cell out of billions! But this is *still* not enough to activate the Helper T Cell. A second signal is necessary, communicated by another set of receptors on both cells.

This second signal is like a gentle kiss from the Dendritic Cell, if you want. It's another confirmation signal that clearly communicates again: "This is real, you are really properly activated!" Why is that so important that we mention it here? This is another security mechanism that prevents Helper T Cells from activating accidentally—only if a Dendritic Cell, that is representing the Innate Immune System here, is activated by real danger is the Adaptive Immune System, represented by the Helper T Cell, supposed to activate.

Let us summarize one last time because this stuff is really important and really hard: To activate your Adaptive Immune System, a Dendritic Cell needs to kill enemies and rip them into pieces called antigens, which you can imagine as wieners. These antigens are put into special molecules, called MHC class II molecules, that you can imagine like hot dog buns.

On the other end, Helper T Cells rearrange gene segments to create a single specific receptor that is able to connect to one specific antigen (a specific wiener). The Dendritic Cell is looking for just the right Helper T Cell that can bind its specific receptor to the antigen.

And if a matching T Cell is found the two cells interlock. But then there needs to be a second signal, like a gentle encouraging kiss on the cheek, that tells the T Cell that all is good and the signal from the presented antigen is real. And only then does a Helper T Cell activate.

OK, phew, wow. Overcomplicated much?

Is this incredibly complicated dance really necessary? Why all these extra steps? Well, to reiterate again: Your Adaptive Immune System is so resource intensive and powerful and, frankly, dangerous to yourself that your immune system *really* wants to absolutely make sure that it does not get activated by accident.

Of course the immune system doesn't want anything as it is not conscious—it's probably more that animals whose Adaptive Immune System could get activated more easily did not survive.

There is another very interesting aspect to the activation of the Adaptive Immune System. In a sense, what is happening here is that information about an infection is transmitted from your Innate Immune System to your Adaptive Immune System.

Earlier we called the Dendritic Cell a living information carrier. By sampling the battlefield and collecting these samples in its receptors, Dendritic Cells become living snapshots of a battlefield in a moment in time. Once one leaves, it stops sampling and is locked in.

After arriving in a Lymph Node, the Dendritic Cell has about a week to find a T Cell to activate before an internal timer runs out and it kills itself, as so many immune cells do. When it does so, it wipes the old information of the battlefield from your body. This wiping of information is another mechanism the Immune System uses to regulate itself. In a sense, the Dendritic Cell is like a paperboy carrying newspapers with breaking news to the Adaptive Immune System.

By sending fresh snapshots, or newspapers, every few hours and eventually deleting them, your immune system collects and delivers a constant stream of fresh information about the battlefield. By deleting it regularly it makes sure not to operate on old information. Today's newspaper with breaking news might carry useful information, while yesterday's is scrap paper, only good to wrap fish.

As the infection dies down no more new Dendritic Cell snapshots are sent to the Adaptive Immune System, the older sets of information die, and no new T Cells are activated. This is a crucial principle that we will encounter over and over: The immune system needs constant stimulation to stay active and by sending living news from the battlefield that self-deletes after a while, your immune system can respond with just the right amount of vigor that is necessary.

Before we move on, here is a really interesting fact: The genes that are responsible for the MHC molecules are the most diverse genes in the human gene pool, leading to a huge variety of MHC molecules between individuals. Of all the things that are different between humans, why are the MHC molecules so unique to every person?

Well, different types of MHC are better or worse at presenting antigens from different enemies, like one type might be especially good at presenting a specific virus antigen while another type could be great at presenting a bacteria antigen. For humans as a species this is enormously beneficial because it makes it really hard for a single pathogen to wipe us out.

For example, when the Black Death ravaged Europe in medieval times, there were people whose MHC class II molecules were just naturally really good at presenting the antigens of the bacterium *Yersinia pestis,* which caused the plague. They had a higher chance of surviving the disease and making sure humans as a species survived.

This is so incredibly crucial for our collective survival that evolution may have made it a contributing factor in mate selection. In human words: You find potential partners with MHC molecules that are different from yours more attractive! OK wait, what? How would you even know this? Well, you can literally smell the difference! The shape of your MHC molecules does influence a number of special molecules that are secreted by your body—which we pick up subconsciously, from the body odor of other people—so you communicate what type of immune system you have through your individual smell!

In German there is even a popular saying "Jemanden gut riechen können," literally translated "being able to smell someone very well," which means "liking someone on an intuitive level." This smell thing is a real thing! Aside from the intuitive level that might feel correct to you, there have been an abundance of studies that showed that all sorts of animals—including humans—prefer the smell of mates with MHC molecules that are different from their own. We just find that if a potential mate has a different immune system, he or she smells sexier. This extra attraction is also a mechanism that avoids inbreeding by making your biological siblings not smell attractive on a sexual level and decreases the chances of close family members getting involved with each other. Which makes sense—by combining genes that create a diverse immune system, the chances of having healthy offspring rise immensely. So next time when you hug your partner, know that their immune system is probably one of the reasons why you find them attractive!

With all this in mind, it's time to finally watch your immune system superweapons in action.

121

20 Awakening the Adaptive Immune System: T Cells

THE AWAKENING OF THE ADAPTIVE IMMUNE SYSTEM USUALLY BEGINS IN the lymph node dating pools, where Dendritic Cells covered in hot dog buns filled with antigens try to find the right T Cells. T Cells have a much more varied set of jobs than the Macrophages or Neutrophils we got to know more intimately earlier. For one, there are multiple classes of T Cells: Helper T Cells, Killer T Cells, and Regulatory T Cells, each able to specialize even more into various subclasses, for every possible kind of infection.*

Looking at a T Cell, you would not be very impressed. They are average size, and don't seem special in any way. But they are absolutely indispensable for your survival. People who don't have enough T Cells, because of a genetic defect, chemotherapy, or a disease like AIDS, have a very high chance of dying from infections and cancers. Sadly, even with the best our modern medicine has to offer, the lives of patients without T Cells often can't be saved. Because as we will learn in a moment, T Cells are the coordinators of

---

* If you have ever played Dungeons & Dragons you may have encountered the same principle of classes before. When you make your character, there are different classes you can choose from, say, a fighter, a mage, or a cleric. But these classes split again into subclasses. For example, a fighter can specialize and become a knight, or a battle master, or a champion (and so on, there are many more). Each of these subclasses still is a fighter, so it smashes heads in with melee weapons, but also has different specialties that make them stronger in different situations. So without the need to create entirely new classes, these subclasses provide a lot more diversity and options for you as the player.

And this is exactly how your immune system behaves. Basically, most immune cells have a number of subclasses with different jobs and specializations and scientists are regularly discovering new ones. For us it is not necessary to learn about each subclass, from Th1 to Th17, it's too complicated and often the differences are very subtle. Like a knight using a sword and a champion using a spear. In the end both subclasses stab monsters with sharp things until they stop moving. We will only mention specific subclasses when they are important enough to do so.

T Cell Career

Pre T Cell

Thymus Training

Virgin T Cell

Regulatory T Cell

Activation through MHC II

Activation through MHC I

Helper T Cell

Killer T Cell

Infection over

Tissue Resident Memory T Cells

Effector Memory T Cell

Central Memory T Cell

the immune system. They orchestrate others and directly activate your heaviest weapons.

T Cells are travelers that start their lives out in the bone marrow, where they mix and match the gene fragments that create their unique T Cell receptors, before they visit the Murder University of the Thymus to be educated. If the T Cells survive their education they move through your lymphatic megacity network, looking for an antigen that is exactly right and to get the encouraging kiss from a Dendritic Cell, to get activated.

You might still think the fact that this principle actually works is a bit crazy. After all, what are the odds that a Dendritic Cell carrying a specific antigen, will find exactly the right T Cell that has the matching receptor for a specific enemy? What are the odds of picking a random puzzle piece out of millions and finding the one cell out of billions that carries the matching puzzle piece that just so happens to fit perfectly into it?

Well for one, it is not just a single Dendritic Cell, in an infection at least dozens will make the trip. And on top of that the system is helped by fast travel. T Cells traverse the whole of your lymphatic superhighway once per day—imagine what this would mean on your human scale. You would need to drive from New York to L.A. every single day, while stopping in hundreds of towns and rest stops on your way to ask around if someone was looking specifically for you. This is what T Cells do and so the chances of meeting exactly the right Dendritic Cell with the matching antigen to their T Cell receptor works out just fine. When this meeting happens the T Cell activates, and all hell breaks loose.

For now we will just talk about the **Helper T Cell** to keep things nice and simple but we will get to know the classes of T Cells much more intimately later on. We talked about the Helper T Cell already but now we are going to get a fuller picture.

Let us think back to our infection. About a day after the Dendritic Cell has left the battlefield, millions of Neutrophils and Macrophages are fighting and dying in a dramatic fashion. At this point there might be just a single activated Helper T Cell in one of your lymph nodes. This is the state of the Adaptive Immune System and somehow it now needs to take control over the situation.

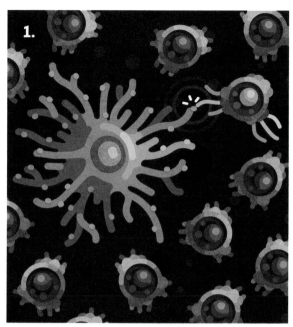

1.

Dendritic Cell presents antigen (Wiener) and looks for a T Cell with matching receptors.

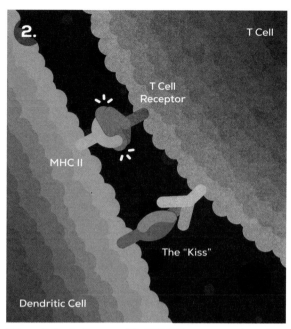

2.

T Cell

T Cell Receptor

MHC II

The "Kiss"

Dendritic Cell

When it finds the specific T Cell they connect and share another signal via a different set of receptors (The Kiss). The Helper T Cell is activated!

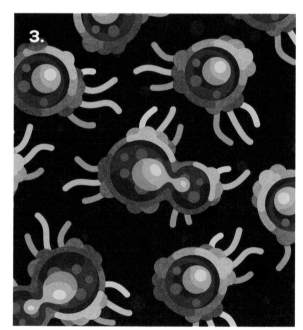

3.

The activated Helper T Cell multiplies rapidly in the Lymph Node and splits in two groups.

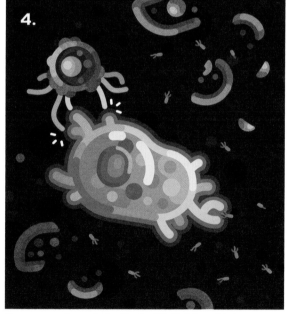

4.

One group travels to the battlefield and takes over command. They put Macrophages into killer mode and decide when the battle is over.

The Helper T Cell can't stay alone if it wants to help beat back the infection, so its first job is to make more of itself. What we will pretty casually describe in the next two chapters is called the *Clonal Selection Theory*. Its discovery won a Nobel Prize and it is one of the most crucial principles of how your immune system works. The theory basically goes like this:

Your activated T Cell leaves the Dendritic Cell that activated it behind and wanders to a different part of the Lymph Node City where it begins the process of cloning itself. It divides over and over again, multiplying as fast as it can. One activated Helper T Cell becomes two, two become four, four become eight, and so on. Within hours there are thousands of them. (And because each of the clones has the same unique T Cell receptor like the first Helper T Cell that got activated, your immune system now has thousands of cells with this unique receptor that is an exact fit to the enemy.)

This growth is so rapid that all the new Helper T Cells begin to crowd out the section of your lymph node megacity.

Once enough clones have been made, the individual cells split into two groups: Let us follow the first group right now! They need a moment to orient themselves and take a deep sniff of the cytokines and danger signals that have been carried by the lymph to the lymph node, and then they follow the chemical track to the battlefield as quickly as they can.

Around five days to a week after the wound was inflicted Helper T Cells arrive at the site of infection, where they begin to act as local commanders. Although Helper T Cells don't do any active fighting themselves, they considerably boost the fighting ability of the local defense cells, specifically of your heavy hitters. For one, they release important cytokines that have a diverse set of jobs, from calling for more reinforcements to increasing inflammation. But Helper T Cells also contribute more directly to the battle by improving the fighting ability of your soldiers. We saw what they do earlier: With a whisper to the black rhino they put them into a wild fighting frenzy, a state of anger that the Macrophage can achieve only with the help of Helper T Cells.

Which makes sense if you think about it—Macrophages are powerful and dangerous monsters and the decision to fully unleash their might should be the result of careful consideration. If they got into a wild battle frenzy whenever a few bacteria showed up, they could do serious harm to the body.

But if Helper T Cells order them to become really properly angry, that means the infection was so serious that the Adaptive Immune System awoke and this permits the Innate Immune System to unleash its full potential. So Helper T Cell commanders at the site of an infection play the role of amplifiers that use the inherent power of the Innate Immune System to overcome harsh enemies.

But Helper T Cells do not just put the Macrophages into Killer mode. Once this battle frenzy is triggered, they are necessary to keep them alive. Helper T Cells monitor the battlefield, and as long as they sense danger, they are stimulated and know fighting is still necessary. Macrophages that fight in battle-frenzy mode are on a timer and will kill themselves after it runs out. This is another one of these safety mechanisms to make sure the immune system is limited to a degree. Helper T Cells can reset this Macrophage suicide timer over and over again. So as long as danger is present, they tell your exhausted warriors to carry on by restimulating them over and over again.

Until they decide to stop doing that. Once Helper T Cells notice that the immune system is clearly winning the fight they stop and so, bit by bit, more and more spent soldiers end their lives. Helper T Cells do not just crank up the violence, they also determine when enough is enough and everybody should calm down.

When the battle is won, the last thing most Helper T Cells do at the battlefield is to kill themselves, joining most of the soldiers in their self-destruction to protect the body from themselves. Except, a few don't. Some Helper T Cells become **Memory Helper T Cells.** Whenever you hear that you are immune to a disease, this is what this means. It means that you have living memory cells that remember a specific enemy. And that enemy might come back, so they stick around and become powerful guardians. Memory Cells are able to recognize a familiar enemy much quicker than the Innate Immune System ever could. In case of another infection, this makes the long trip of the Dendritic Cell to the lymph node unnecessary because these Memory Helper T Cells can immediately activate and call for heavy reinforcements.

This memory reaction is so fast and so brutally efficient that most pathogens get only a single chance to infect you. Because your Adaptive Immune System has adapted and remembers. But Memory Cells get their own chapter later on so we'll stop talking about them for now.

The importance of the Helper T Cell does not stop here, not even close. Remember, we followed only one group from the lymph node to the battlefield. There was a second group that remained and what they are about to do might be even more important: Activating some of your most efficient immune weapons you have at your disposal. The mighty **B Cell**, your living weapon factories.

21 Weapon Factories and Sniper Rifles: B Cells and Antibodies

B Cells are large, blob-like fellows that share a few characteristics and properties with T Cells, namely that they originate in the bone marrow and that they have to undergo the same brutal and deadly education—only it doesn't happen in the Thymus but directly in the bone marrow.*

Just like their T Cell buddies, all your B Cells combined come with at least *hundreds of millions to billions of different receptors for millions of different antigens.* And just like T Cells, *every single individual B Cell has one specific receptor that is able to recognize one specific antigen.*

What makes B Cells special, and very dangerous for friends and foes, is that they produce the most potent and specialized weapon the immune system has at its disposal: Antibodies. Antibodies are weird things and pretty complex and fascinating, so we will brush over them here and discuss them in the detail they deserve a bit later, but in a nutshell—Antibodies are basically *B Cell receptors.* Antibodies themselves are a bit like crab-like sniper rifles, as they have been made against a specific antigen—and therefore a

---

* Do you think the "B" in the name "B Cell" was chosen since B Cells originate from the bone marrow, because the T Cell has its "T" from the Thymus? Well sorry, no, that is just a coincidence and would make way too much sense to fit into the mess that is the language of immunology. The "B" in B Cells comes from the "Bursa of Fabricius," which is a saclike mini organ that sits right above the end of the gut in birds. This organ had been known for hundreds of years but nobody had an idea what it was doing. Until a graduate student did some work with chickens who were missing their bursas and subsequently found that they were unable to produce Antibodies. He discovered B Cells, the factories that produce Antibodies, and that they were made in this weird little organ in birds, which was a huge breakthrough in immunology and created an entire new field of study. Humans don't have a bursa, we use our bone marrow to make B Cells. But yeah, while the name makes sense, it still is a missed opportunity.

specific enemy—so they in the metaphorical sense hit a pathogen right between the eyes.

OK wait, how can something be a receptor and a weapon that floats around at the same time? So basically, Antibodies are stuck to the surface of B Cells and work as their B Cell receptors, which means that they can stick to an antigen and activate the cell. Once a B Cell is activated it begins to produce thousands of new Antibodies and starts to vomit them out, so they can attack your enemies—up to 2,000 per second. All Antibodies are produced this way. But they'll get the love and attention they need after we finish explaining the B Cells that produce them. For now remember only one thing: Antibodies are B Cell receptors that are vomited out by the cell by the thousands per second when they are activated!

Before we go on, a short disclaimer. B Cell activation and their life cycle are complicated. Many things we learned about come together here, a lot of stuff is going on simultaneously as there are a lot of parts of the immune system that become intensely intertwined. So while you read the next few paragraphs you may think "Phew, this is a lot to take in." Don't worry, we will take breaks and summarize and solidify what we learn in this chapter.

This is the most complex process we will describe in this book, so we'll take it slowly and one step after another. The payoff is really worth it because once you roughly understand this layer of complexity, even on a surface level, you can really appreciate how seriously breathtaking your immune system is. Also afterwards the rest of the book is smooth sailing.

OK, let's get on with it! So as we said in the beginning B Cells are born in your bone marrow where they mix and recombine the gene segments responsible for their *B Cell Receptors* to be able to connect to one specific antigen (if you think back to our metaphor with cooking a lot of tasty dishes, each B Cell with its specific receptors represents one dish). After they did that, similarly to T Cells, they have to undergo a harsh and deadly education to make sure that they are not able to connect their unique receptors to the proteins and molecules of your own body. The survivors become traveling virgin B Cells, inactive cells that move through your lymphatic system every day, just like T Cells, making the trip from New York to L.A., stopping in hundreds of towns and rest stops to check if someone is looking for them. But this is where the similarities between T and B Cells end.

B Cell Career

Pre B Cell

Bone Marrow
Training

Virgin B Cell

Activation #1
through Antigen

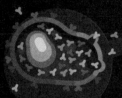

B Cell

Activation #2
through T Cell

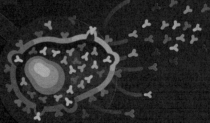

Plasma Cell

Infection over

Infection over

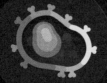

Memory B Cell

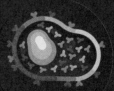

Long Lived
Plasma Cell

In the lymph node megacities there are specific B Cell areas where they hang out for a bit and have a coffee and a chat, waiting a bit to see if they are needed. B Cells are very dangerous cells, so they need a strict two-factor authentication to be truly activated—one by the Innate Immune System and another one by the Adaptive Immune System!

We'll break this down into individual steps and summarize them at the end.

Step One: B Cell Activation by the Innate Immune System

To understand the first step we need to think back to the infrastructure of your immune system and how it is connected. Let's remember the infection of your toe, where a massive battle has been going on between your Macrophages and Neutrophils and the bacteria that infected your flesh, for maybe a day or two.

This battle has not been without casualties, and lots and lots of bacteria have been killed. Many of them have been swallowed whole by your Macrophages but this is not all. Many others were ripped apart by the dangerous weapons of Neutrophils, they got bled out by complement proteins (the invisible army) ripping holes into them, or they were torn apart trying to escape a Neutrophil NET (if you forgot what these were, this was the part where Neutrophils basically exploded their DNA, spiked with harmful chemicals, to create barriers around them to trap pathogens). Just through the sheer violent efforts of the immune reactions, a lot of death was dished out.

Given enough time, eventually your immune cells will clean up but for now they are more concerned with killing and fighting bacteria that are still alive. So the battlefield is filled with death and suffering. A considerable number of bacteria pieces and carcasses float around at the site of infection, many of them covered by complement proteins. It is really like a war where combatants fight knee-deep in the bloody and ripped-apart bodies of their enemies and friends.

But already the ingenious mechanisms of your immune system infrastructure begin to clean up and filter. As we mentioned before, inflammation ordered by the immune cells and caused by other dying cells diverts a lot of fluid from your blood to an infection, which floods the battlefield considerably. The longer the fighting is going on, the more fluid is coming in. But this can't go on forever or your tissue would burst, so some of the fluid also has to leave the site of infection again.

We did learn before what your body does with excess fluids in your tissue, it consistently washes them away, right into the Lymphatic System. The fluid and with it a lot of the battlefield detritus, with parts of dead bacteria and spent cytokines and other garbage, become part of your *Lymph*. Remember lymph is this weird, slightly disgusting fluid that is constantly collected from all the tissues in your body. And in the case of an infection, the lymph is carrying all the dead and dismantled bacteria with it, a lot of them covered in complement proteins. This way the lymph flowing through you is a *liquid information carrier*.

And this information is headed towards the next immune system base, the megacities and intelligence centers of the lymph nodes. Once it arrives here it is drained through the B Cell area where thousands of virgin B Cells hang out. The B Cells get right in the middle of the stream of fluid information and let the lymph flow around them and their B Cell receptors, which sift and explore all the antigens and detritus coming from your tissue.

The virgin B Cells look specifically for antigens they can connect to with their special and unique B Cell receptors. They are fishing for the one antigen they can connect to, so they know they can activate!

OK, so far so good, but you may have noticed something: There is no Dendritic Cell involved here, so does this mean B Cells don't need to go through this dance with another cell? It all has to do with a huge difference between T Cell receptors and B Cell receptors and it is important enough that we will explain it right now. Let's do this with sausages again!

Remember the MHC class II molecule? The hot dog bun that presented an antigen, the wiener, to the T Cell receptors, so it could activate? T Cell receptors are really picky eaters that eat only wieners and only if they come inside a bun. But this has a major consequence for T Cells: The antigens that can activate T Cell receptors have to be pretty short because the MHC

133

molecule can only carry short antigens. The hot dog bun on the Dendritic Cell can only hold wieners. In contrast, the B Cell receptors are not as picky.

Both T and B Cell receptors are each made to recognize a specific antigen, but your B Cells are far less restricted. So T and B Cells recognize very different things in size and dimension. Not only can B Cells pick antigens right from the fluids around them and activate, they also can sort of pick up a much larger piece of meat, to get back into our food metaphor.

Wieners are highly processed meats that don't have a strong resemblance to the animal parts that they are made of. And so are the antigens T Cells can recognize. The antigens B Cell receptors can recognize are a bit like huge roasted turkey legs, with bone and skin. T Cells are way too picky for that, B Cells don't care.

And B Cells do not need an MHC molecule, they don't need to get a presentation from another cell like T Cells do. No, B Cells can pick up large antigen chunks (the turkey drumsticks) directly from the lymph that flows through your lymph nodes.

OK, so we learned two things now: Virgin B Cells sit in your lymph nodes, where they bathe in lymph and take in all the antigens that are transported through the area from the closest battlefield. Their B Cell Receptors can just grab big chunks of antigens directly from the lymph and this way B Cells can get activated.

But there is more: B Cells do have more direct help from the Innate Immune System. Did you think it was suspicious that we kept mentioning that the bacteria from the battlefield were covered in complement proteins? B Cells cannot just recognize the dead bacteria antigens, they also have special receptors that are able to recognize complement proteins.

We mentioned before that the Innate Immune System is responsible for activating and providing context to the Adaptive Immune System and here we are encountering this principle once more! By being attached to the pathogens, the complement system is officially confirming to the B Cell that there is a real danger. So complement proteins attached to an antigen makes it about 100 times easier to activate a B Cell than it would be to activate it without the complement. This multilayered complexity of parts interacting so elegantly and communicating with so much care is one of the things that make the immune system so beautiful and amazing. (You can imagine complement proteins on an antigen as a really nice sauce on the turkey leg, which makes it even tastier to your B Cells.)

Fun fact, this was just step one of the B Cell activation, but it is already incredibly important because it will trigger a quick reaction to an infection. Without any extra steps, these simple mechanisms that happen on their own because the lymphatic system is always draining your tissue, are creating a relatively fast response. This is especially important in the early stages of an infection, when not that many Dendritic Cells have reached the lymph nodes to activate Helper T Cells.

OK, take a short breather and revisit what we just learned: Battlefield, dead bacteria covered in complement, lymph taking away these carcasses, B Cells inside the lymph node pick it up, and now finally: Early B Cell activation!

What does this early activation look like? Well, first of all the activated B Cell moves to another area in the lymph node and begins cloning itself. One becomes two, two become four, four become eight, and so on. This cloning continues until there are roughly 20,000 identical clones, all with copies of the specific receptor that was able to connect to the original antigen, the first virgin B Cells picked up. These B Cell clones begin producing Antibodies that use the blood as a lift to the site of infection and can flood the battlefield and help out—although they are second-rate antibodies. They are OK at their job but not amazing, snipers that land more body shots than head shots.

Without a second step, without activation number two, most of these B Cell clones will kill themselves within a day. Which makes a lot of sense actually because if they don't get activated again, these B Cells have to assume the infection was pretty mild and they are not actually needed that badly—so to not waste any resources or do any needless damage, they will kill themselves.

To be truly awakened we need the second part of the two-factor authentication. And this one is provided to the B Cells by their colleagues from the Adaptive Immune System, or more precisely, by activated Helper T Cells.

Step Two: B Cell Activation by the Adaptive Immune System

As we learned in the last chapter, after a Helper T Cell has been activated and has made a lot of clones of itself, one group of the Helper T Cells moves to the battlefield while the other group goes off to activate B Cells for real.

In a nutshell, an activated T Cell needs to find an activated B Cell and BOTH cells need to be able to recognize the same antigen! OK, wait a second. So are we seriously saying that two cells in your body mix gene fragments randomly, with hundreds of millions to billions of possible outcomes? And then a pathogen shows up and coincidentally both need to be activated independently and then they need to meet each other? And only then, in this absurdly specific and seemingly impossibly unlikely case will your immune response fully be activated? Well yes, although the way this works is a bit mind-bending and the fact that nature came up with it is extremely elegant.

Basically, in order to be properly activated, B Cells have to become *antigen-presenting cells*. This works because B Cell Receptors are very different from T Cell receptors, which need the hot dog bun to recognize a very tiny piece of antigen. Picky eater vs not picky eater, remember?

So when a B Cell receptor connects to a turkey drumstick, a big chunk of antigen, it swallows it and processes it inside itself, just like a Dendritic Cell would. It slices the big chunk of meat up into dozens or even hundreds of tiny sausage parts, all the size of wieners. And these tiny parts are then put into MHC molecules (hot dog buns) on the surface of the B Cell. *Basically, a B Cell takes a complex antigen and turns it into many processed, simpler pieces that are then presented to the Helper T Cell.*

Think about what the immune system is doing here, it increases the chance that a B Cell and a T Cell are able to match with each other *massively*. The B Cell is not just presenting a single specific antigen. It is presenting dozens or even hundreds of different ones in its MHC molecules! Hundreds of different wiener-sized sausage pieces in hundreds of different hot dog buns. *So technically, the B and T Cells don't recognize the exact same antigen.* This is good enough for the Adaptive Immune System because this means if a Helper T Cell can connect to the antigen presented by a B Cell, there is an enemy out there and both cells are able to recognize it. This is the secret of B Cell activation: B Cells can ONLY be fully activated through the two-factor authentication.

OK, stop! All of this is a lot of information.

And if your head is smoking and your eyes are spinning right now, this is actually the correct reaction. There are a lot of things going on, over a long amount of time, in many different places and a lot of different cells. So you are in good company if you are confused and it is time to summarize what happened here.

Step 1: A battle needs to occur and dead enemies, which are big chunks of antigens (turkey drumsticks), need to float through the lymph node. Here, a B Cell, with *a specific receptor* needs to connect to the antigen. If the dead enemy is covered in complement, activation will be much easier. This will activate the B Cell, which makes a lot of copies of itself and produces low-grade antibodies, but the B Cells will die after around a day if nothing more happens.

Step 2: In the meantime, a Dendritic Cell needs to pick up enemies at the battlefield and turn them into antigens (wieners) which are put in the MHC class II molecules (hot dog buns) and travel to the T Cell dating area in the lymph node. Here it needs to find a Helper T Cell that is able to recognize the antigen with its unique T Cell receptor (eat the wiener from the bun). If this happens the Helper T Cell is activated and makes a lot of copies of itself.

Step 3: The B Cell breaks the big chunk of antigen (turkey drumstick) down into dozens or hundreds of small antigens (wiener sized) and begins presenting them in MHC class II molecules (hot dog buns).

Step 4: An activated B Cell that is presenting hundreds of different antigens (wiener-sized sausage pieces) needs to meet a T Cell that can recognize one of these antigens with its specific T Cell receptor, which is the second signal for the B Cell.

Only if this exact sequence of events occurs does a B Cell get activated for real. Are you impressed yet with your biology?*

---

* Funny story: This is actually still simplified and we are leaving out a few major details that are actually pretty important. We'll address some of them at various places in the book. But honestly, this stuff is mind-bendingly unintuitive and hard even in a strongly simplified form. If you manage

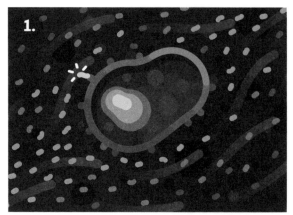

1. Antigens from a battlefield float through the lymph node, where a Virgin B Cell connects to it.

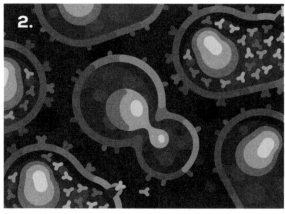

2. This will low-key activate the B Cell, which makes a lot of copies of itself.

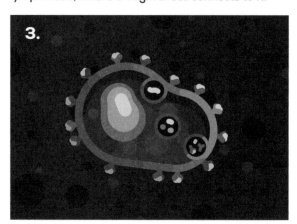

3. The B Cell breaks down the antigens into smaller ones and presents them in MHC class II molecules.

4. Meanwhile Dendritic Cells pick up antigens, present them in MHC class II molecules, and activate matching T Cells.

5. B Cell meets the activated T Cell that can recognize one of these antigens with its specific T Cell receptor.

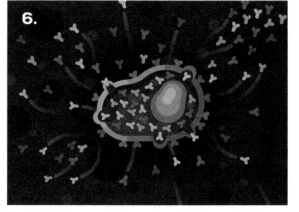

6. The B Cell is now a fully activated Plasma Cell!

Can you appreciate the level of sophistication that is happening here? How insane it is to make billions of individual T and B Cells, activate them individually through different paths, and then expect them to meet each other? Evolution and time are truly incredible at crafting insanely complex and elegant mechanisms. If this sequence of events happens, the last and most powerful stage of the Adaptive Immune System finally starts up in earnest and awakens. Now all the conditions the immune system could ever ask for have been fulfilled. It now knows for sure that there are a lot of enemies active inside the body.

Your B Cell that was properly activated through the two-factor authentication now changes. It has waited its whole life for this moment. It begins to swell, to almost double its size, and transforms into its final form: the **Plasma Cell**.*

The Plasma Cell now begins producing antibodies for real. It can release up to 2,000 antibodies per second that saturate the lymph and blood and the fluids between your tissue. Like the Soviet rocket batteries in World War II that could send a never-ending barrage of missiles on enemy positions, antibodies are made in the millions and become every enemy's worst nightmare, from bacteria to viruses or parasites. Even cancer cells. Or, if you are unlucky and have an autoimmune disease, your own cells.

OK. Phew. What a complicated affair. But wait, there is more, one last aspect of B Cell activation that makes this process even more genius. Now your immune system truly begins beating microbes in their own game as a beautiful dance begins, a dance that makes your defenses even better and stronger.

---

to remember that B Cells activate by picking up stuff themselves and then get activated a second time by T Cells, that is already amazing. You don't need to remember any more details than that to know an impressive amount about your immune system! But this stuff is just too cool to not try to get this magic across.

* If you happen to be just the right age this might mean something to you: In a sense, B Cells are Saiya-jin and Plasma Cells are Super Saiya-jin. For everyone who never watched *Dragon Ball Z,* this is just a charming way of saying that B Cells are strong fighters and Plasma Cells extremely strong fighters, also possibly blond with a lot of product in their hair. Let's end this footnote before it becomes more embarrassing.

22 The Dance of the T and the B

ONE THING THAT WE ELEGANTLY AVOIDED MENTIONING SO FAR IS HOW good the B Cell receptors are at recognizing antigens. Earlier we described these receptors and antigens as puzzle pieces fitting together perfectly. Well, that was sort of a lie, sorry. As it is said, perfect is the enemy of good, and during a dangerous infection your immune system has no time to wait for the *perfect* match—it really is cool with a good or even OK-ish match. So B Cells can get activated if their receptors are *good enough* at recognizing an antigen.

The immune system evolved this way because it is better to have *some* working weapons as quickly as possible than perfect weapons after all the damage is already done. But this also weakens your immune defense. As we said, on the level of proteins, *shape* is everything, and having Antibodies with a really good shape that fits really well on an antigen is an incredible advantage that can mean the difference between life and death. And your immune system wants it all, a quick response and then a perfect defense.

So your immune system came up with a way to produce OK-ish Antibodies as fast as possible but also with an ingenious system to fine-tune and improve the Antibodies, so they become actually perfect weapons against the antigen. It all starts with a dance.

We said before that B Cells need to be activated by Helper T Cells (that were activated themselves by Dendritic Cells) to turn into Plasma Cells but actually this process is a tad more amazing and sophisticated. Your immune system makes sure that only B Cells that are able to make *amazing* Antibodies turn into Plasma Cells. OK, so how does this work?

Well to be honest it is a bit of a hot mess so we will simplify a bit here. In a nutshell, if a T Cell recognizes an antigen that a B Cell presents to them, it stimulates the B Cell. This stimulation is like a gentle kiss or a warm encouraging hug. Not only does this prolong the life of the B Cell, it also motivates it to try to improve the Antibody!

Every time a B Cell receives a positive signal from a Helper T Cell, it begins a round of purposeful mutation. This process is called *Somatic Hypermutation* (also known as *Affinity Maturation*) and we will never use this clunky term again.

Like a cook that works and improves a recipe that got praise from food critics, the B Cell begins to enhance and refine the recipe. The B Cell *mutates* the gene fragments that make its receptors, and therefore its antibodies.

What B Cells do here is basically go back into the kitchen during the dinner party. At this point the guests have shown up and they know what sort of dinner they want to eat. So now they begin to randomly change the recipes a little bit here and there. The goal is to make just the *perfect* dish, just like a three-star Michelin restaurant. Not good, not great, perfect! So maybe in the original recipe, the carrots were finely chopped and the beef was roasted. Now the B Cell might cut the carrots into sticks and grill the beef instead. No new ingredients but fine-tuning the way they come together to create the final dish.

The goal is to engineer the perfect dinner for the guests, something so good they will be ecstatic. The perfect Antibody for the pathogens. But how do your B Cell chefs find out if the guests like the enhanced recipe more than the original one, if their new Antibody is a better fit than the original one? Well, exactly the same way the B Cells were activated in the first place: They basically bathe their new and improved receptors in the stream of lymph from the battlefield that is passing through the lymph node. If a battle is still ongoing there should be plenty of antigen passing through.

If the random mutation (the fine-tuning of the dish) made the B Cell receptor worse, then it will have a harder time picking up antigens. It will not get stimulation and kisses from T Cells. Which will make it sad and after a while it will kill itself.

But if the mutation improved the B Cell receptor it will now be even better at recognizing the antigen and the B Cell will get an activation signal once again! Once this happens, it takes in the chunk of antigen (the turkey drumstick) and cuts it into many smaller pieces (the wieners) and once again tries to present them to a Helper T Cell. You can imagine this as if the B Cell chef is super excited and happy about their improved recipe and wants to tell the world about it!

In our cooking metaphor the Helper T Cells might be a food critic coming from the dining room and showering the B Cell in praise and kisses. And this encouragement motivates your B Cell chefs to improve the dishes even more! And the cycle repeats.

Over time a natural selection takes place. The better B Cell receptors become at recognizing the antigen that flows through the lymph node, the more stimulation and encouragement they get. While at the same time, the B Cells that get worse or don't improve kill themselves.

In the end only the best possible B Cells survive and go on to make many new clones of themselves! These are the B Cells that eventually turn into Plasma Cells, that fine-tuned their receptors and are able to make the best possible weapons against the enemy. This is the reason Antibodies are so deadly effective, why they hit an enemy between the eyes like a sniper. They were not just randomly chosen, they were molded and improved and fine-tuned until they were perfect. This is the reason that even if you know nothing about the immune system, you have probably heard the term "Antibody" a bunch of times from medical professionals. They are your superweapons, the main reason you can survive serious infections.

This mechanism makes the Adaptive Immune System actually *adapt* to the enemy in real time. We asked before how you could keep up with the billions of different enemies that are also able to change themselves. *This is one way.* A system that can replicate very quickly, that has a defined target and can adapt to it fast, that fine-tunes and improves its weapons until they are perfect. What a beautiful and ingenious solution that shows that the Adaptive Immune System really deserves its name—it truly can beat microbes at their own game.

If you made it through the last two chapters—great job. And I mean that, this stuff is not easy and believe it or not, this is the simplified version. The immune system and the universe in general are unfortunately not made to be intuitively understood by apes with smartphones, and sometimes this makes it really hard to do a deep dive, even if a topic is important. You don't need to remember everything you just read in detail.

Actually, I would wager that it is impossible to exactly recall what you just learned by reading this chapter only once. And that is totally OK. You learned principles and also you made it through the hardest part of the book! This was the complexity peak and from now on it's pretty smooth sailing for the most part! We are also almost back at telling crazy battle stories!

To get the full picture of the immune system we only need to talk about the actual weapons—the sniper rifles!

23 Antibodies

ANTIBODIES ARE AMONG THE BEST AND MOST SPECIALIZED WEAPONS YOUR immune system has at its disposal. Produced by B Cells, Antibodies themselves are not particularly deadly. They are actually nothing more than mindless protein bundles that can stick to antigens. But this they do with extreme efficiency.

You can imagine them as a sort of hashtag of death. The most common Antibodies are shaped like little crabs with two pincers and they are seriously pretty tiny: For an average-sized immune cell, an Antibody is the size that a grain of quinoa is to you. In a sense they are sort of comparable to the proteins of the complement system—which are also nothing more than tiny proteins that float around—but with one huge difference: Complement proteins are generalists, while Antibodies, as we just learned, are specific.

This makes it incredibly hard for a pathogen to hide from Antibodies, since they are made specifically for them. Like a magnet, Antibodies will seek them out and grab their victim with their tiny pincers. And once an Antibody has attached itself, it will not let go again. This is basically what they are: Tiny crab-like proteins that are extremely good at grabbing on to the enemies they were made for, better than anything else the body has to offer because as we briefly mentioned before, Antibodies *are* B Cell receptors.

What makes them so extremely effective is their anatomy. Every Antibody has those two pincers, each of them able to grab a specific antigen extremely firmly. And they have cute little butts that are extremely good at connecting to your immune cells. The pincers are for enemies, the cute butts for friends.

With these tools Antibodies do multiple things: First, similar to complement, they can *opsonize* enemies. In this context it means that Antibodies swarm an enemy and grab them, which makes their victim more delicious for your soldier cells to eat. They grab the pathogen just like an angry crab that pinches you would because you annoyed it. It would be very hard for you to live a happy life if you were covered in wiggling, buzzing tiny crabs that you could never get rid of. Sounds like right out of a horror movie.

When the Antibody army arrived at our infected toe, the bacteria that were covered by them were equally unhappy with their life situation and totally helpless. But Antibodies don't only make pathogens helpless, they also can maim them and make them unable to move. Or in the case of viruses antibodies can directly neutralize them and make them unable to infect cells.*

Even worse, because Antibodies have more than one pincer, they can grab more than one enemy, and if they do, these two are now bound together. If millions of Antibodies flood a battlefield, they can clump up large piles of pathogens together that are now even more helpless and even more unhappy and scared, as a big pile of victims are even easier to detect for your Macrophages and Neutrophils, who gladly swallow them whole or shower them in acid. Imagine that—trying to invade an enemy position and then being bound together with a few dozen of your buddies by tiny pinching crabs. Unable to move or to act, while an enemy soldier comes at you with a crazy laugh and a flamethrower.

And similarly to complement proteins, Antibodies also support your soldiers directly: As you can imagine, bacteria prefer not to be grabbed and tossed into a bath of acid to die horribly. So they evolved to avoid the grip of death by Macrophages and Neutrophils. Bacteria are sort of slippery, like oiled piglets that run around in a panic. Antibodies act as a special superglue— your immune cells, specifically your phagocytes, the cells that eat enemies alive, can grab the butts of antibodies very easily. It is like the difference between trying to open a slippery glass of pickles with wet or with dry hands.

Here another safety layer of your immune system comes in. The cute butts of Antibodies that are for friends are in a sort of "hidden mode" when Antibodies just float around, so immune cells can't just pick them up from

---

* What do we mean when we say that an Antibody "neutralizes" a virus? Well, imagine your cells are a subway train and the virus a passenger that wanted to get inside. This is usually fairly easy for the virus, just pass one of the automated ticket barriers and enter through one of the doors. An Antibody is basically grabbing and covering up the ticket of the virus so it can't pass the ticket barrier and is stuck outside. The more Antibodies attach to the ticket, the more impossible it becomes to get to the train. And so it is neutralized, unable to do anything of consequence. A passenger stranded at the station.

fluids. As soon as an Antibody has grabbed a victim with its tiny pincers, its butt changes its shape and is now able to bind to immune cells. This is pretty important since your body is teeming with antibodies at any time, and it would cause all sorts of chaos if your immune cells would bind to the antibody-butts just randomly.

Another thing Antibodies can do with their cute butts is to activate the complement system. Complement, as efficient and deadly as it is, when it is only by itself, its abilities are limited and it relies on finding the surfaces of enemies with a lot of luck basically. Remember, it just kind of floats passively in the lymph. And some bacteria are able to hide themselves from the complement system so that it doesn't activate close to them. Antibodies are able to activate the complement system and sort of attract it to bacteria, increasing its efficiency wildly. Again we see the principle of our two immune systems: The innate part does the actual fighting, but the adaptive part makes it more efficient with deadly precision.

Antibodies are not just tiny crabs though. There are multiple classes that actually do very different things and are used for different situations. Of course their names are unintuitive and hard to remember—so we will go through them *very* briefly. When we mention them again and their class is important, we will give a short reminder of what their job was again, so technically you can skip the next part if you want to get to our next story.

An Aside The Four Classes of Antibodies*

IgM Antibodies—the First Defenders on Site

IgM Antibodies are usually the majority of Antibodies B Cells produce when they get activated. They were most likely the first Antibodies to evolve hundreds of millions of years ago. IgM is basically five Antibodies merged to-

---

* OK, OK, there are five antibody classes in humans but we are going to ignore the poor IgD Antibody because it is not relevant for anything we are talking about in this book. In a nutshell, IgD can help with activating a bunch of immune cells and whatnot. But I think we had enough details already and this is not that important. But there you have it, a footnote to a headline!

gether at the hips, which has the advantage that they have five butt regions. Two of these butt regions together can activate an additional complement pathway. More activated complement protein means more immune cells drawn to the enemies. Early in an infection this has the upside that while the adaptive immune system is still booting up and is not in full combat mode, IgM Antibodies are already making the innate immune system more deadly and more precise. Especially against viruses, IgM antibodies are a powerful early weapon that can slow down an infection. With their ten pincers they can clump them together easily. So IgM Antibodies are the first Antibodies to be deployed—which also means that they are the least refined through mutation and the dance of the B and T Cells. Which is OK since their most important job is to buy time until better Antibodies are available.*

IgG Antibodies—the Specialists

IgG Antibodies come in a few different types. We don't need to know them in detail, let us just treat them as different flavor variants of the same ice cream. The first IgG flavor is a little bit like complement—it is really good at opsonizing a target and just covering it like an army of fruit flies, making it harder for a bacteria to do its thing and to function properly. Its tiny butts are like special glue that your phagocytes can grab on to easily and devour an enemy with much less resistance. In general IgG is not nearly as good at activating complement as IgM is, but it is still pretty solid at it.

Another one of the flavors of IgG is especially useful if an infection has been going on for a while. If this is the case it is very likely that a plethora of

---

* We mentioned before that your spleen is a sort of lymph node for your blood but there is more! This tiny organ is the main source for superfast responding IgM Antibodies in your blood. A sort of emergency base that can react fast if pathogens like bacteria make it into your blood flow, via an injury for example. The spleen filters your blood and when it finds enemies here it can quickly activate B Cells that rapidly make IgM Antibodies. Sure, they are not optimized like the other Antibody classes, but they are available extremely quickly, which is important when you have in-vaders in your blood—that gives a pathogen access to the whole body! This is one of the things that makes your spleen so important. This mechanism was discovered after wars, where due to severe injuries to the torso, people often had their spleens removed. It turned out that many of them died of sepsis later in life, at much higher rates than the rest of the population. Nowadays, if your spleen is damaged, say in a car crash, doctors try to save as much of it as possible.

immune system actors have created plenty of inflammation already. And as we learned, inflammation, as useful as it is, is not super great for the health of civilian cells and the body in general. Especially if an infection is getting chronic. So these special IgG Antibodies are specifically made to be unable to activate the complement system late in an infection, which limits inflammation.

Another thing that makes IgG Antibodies special is that they are the only Antibodies that are able to pass from the blood of a mother into the blood of an unborn fetus via the placenta.

Not only is this protecting the unborn from a virus infection its mother may suffer from, it does so way beyond birth. IgG is the Antibody that takes the longest to decay, so it gives a newborn human a passive defense against virus infections that protects it in the first few months until its own immune system has a chance to properly boot up by itself.

IgA—Making Poop and Protecting Babies

IgA is the most abundant antibody in your body and its main job is to serve as a cleanup mechanism for your mucosa. Or in other words, it is in abundance in your respiratory tract, your primary sexual organs, and especially in your digestive tract, including your mouth. Here, a great number of special B Cells produce large amounts of these special antibodies. IgA is basically a sort of bouncer that protects the entrance doors to your insides, your eyes, nose, mouth, etc. from unwanted guests, by neutralizing pathogens early on before they have the chance to get in and establish a foothold.

They are the only antibodies that can freely pass the internal border of the Mucosa Kingdom from the inside to saturate our mucosa on the outside. So if you have a nasty cold, your snot is full of IgA giving viruses and bacteria a hard time.

IgA is different from other antibodies in one major way: IgA have their little butts merged together, which means that IgA can't activate the complement system at all. This is no accident: An activated complement system means inflammation. And since IgA Antibodies are constantly produced in your gut, if they could activate complement this would mean that your gut would be constantly inflamed. Which would cause disease and diarrhea and

Antibodies

Antibodies themselves are not particularly deadly. They are actually nothing more than mindless protein bundles that can stick to antigens. You can imagine them as a sort of hashtag of death.

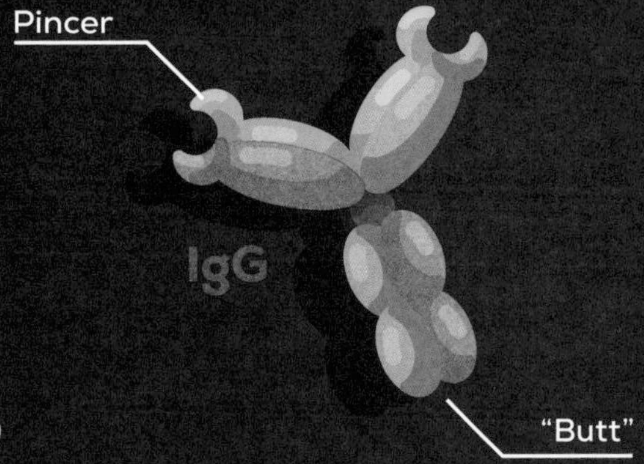

Pincer

IgG

"Butt"

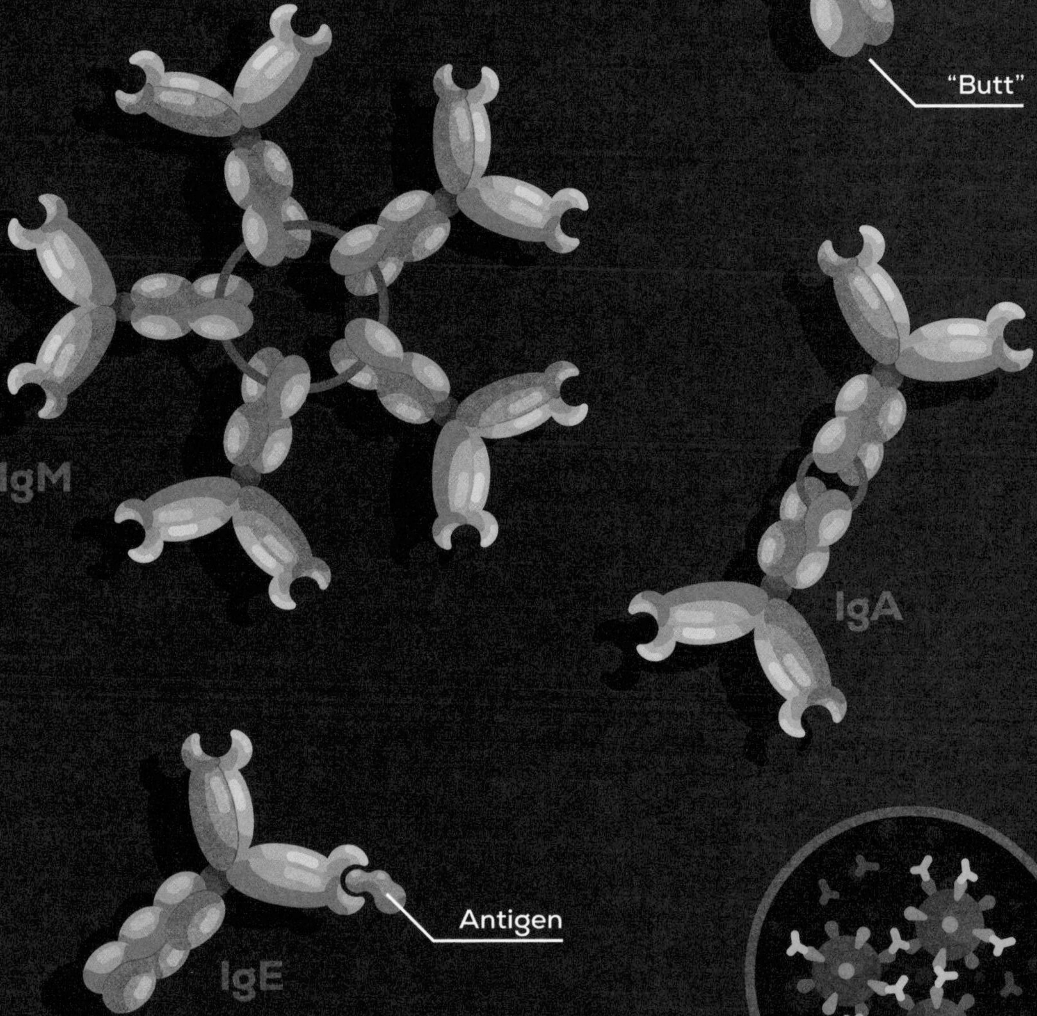

IgM

IgA

IgE

Antigen

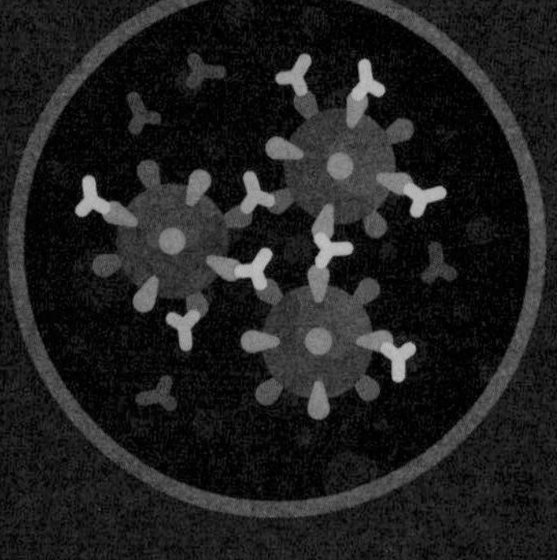

Antibodies (yellow) clump up viruses

make you very unhappy. Diseases that cause constant inflammation in the gut region, like Crohn's disease, for example, are no joke and can seriously impede the happiness and well-being of a patient suffering from it.

One of the things IgA is great at is attacking multiple targets and clumping them together in chunks of really unhappy bacteria that are then swept away by snot, mucus, or your feces. Up to a third of your feces are actually bacteria unlucky enough to get caught up by the poo on its way out. Once they are on board, there is no way to get out again. Other than protecting and cleaning up your gut, IgA also protects our babies. When mothers are breastfeeding they provide their offspring with a large amount of IgA Antibodies through their breast milk. These antibodies then cover the gut of the newborn and protect its still-fragile intestinal tract from infections.

IgE Antibodies—Thanks, I Hate It

IgE Antibodies don't look that special, to be honest, but you can imagine them as giving you the finger with their two little pincers if you want. If you ever had the very unpleasant experience of suffering an allergic shock, you can thank the IgE Antibody for the amazing time you had that day. Or in a less life-threatening scenario, they are the thing that makes you have allergic reactions to things that are harmless. From the pollen of plants to peanuts or the sting of a bee. Of course, evolution did not just come up with the concept of allergic reactions to mess with you for no reason. The original purpose IgE Antibodies serve is to protect you against infections by huge enemies: Parasites. Especially worms. The how and why is a story that is best told in its own chapter, so for now, let us ignore IgE Antibodies and allergies and shake our fists angrily at them.

How Do B Cells Know What Sort of Antibody to Make?

Now you may ask yourself: If there are so many different types and variants, how do B Cells know what sort of antibody is needed? After all, the different classes of Antibodies do very different jobs very well but are pretty useless at others.

We said before that the Dendritic Cells carry snapshots from the battlefield to provide context. This snapshot of the context at the site of infection is then communicated to the Helper T Cell. As time moves on, new Dendritic Cell snapshots with different contexts arrive from the battlefield. And so what was right at some point of an infection might change over time.

So B Cells are not locked into making a certain class of Antibody—they always start with IgM but can switch the Antibody type if the Helper T Cell asks and encourages them to! Having a nasty cold or a gut infection and need a lot of antibodies in your snot or stool? Make IgA! Having a parasitic worm in your intestines? Make IgE! A lot of bacteria have infected a wound? Make IgG flavor one! There are a lot of virus-infected cells? Please, more IgG flavor three! (Once an Antibody class has been switched there is no going back though.)

The amazing ability to collect and communicate intel on this level of ingenuity is another testament to the stunning brilliance and beauty of the great concert of the immune system. All the parts working together, changing and working and coordinating, without any single part being conscious or aware.

OK! You are done with the first section of the book! You learned so much about so many different parts of yourself! You also finished the hardest part of the book! Let's take a big step back for a moment to reflect on what we've learned so far.

We learned about the scope of your body, your cells, and some of your most common enemies, bacteria. About your soldier and guard cells that guard your insides, the mechanisms they use to identify and kill invaders, and how they use inflammation to prepare the battlefields of your body. We learned how your cells recognize things and how they communicate with each other. We explored the complement system that saturates every fluid in your body. We learned about your surveillance cells that get help when necessary. We learned about your internal infrastructure and how your body has billions of different weapons made by recombination, how these superweapons are deployed and improved upon through mutation. And of course we learned about your first line of defense, your skin, and what a hellhole it is.

But if you think about it, compared to other diseases, how often do you hear about people getting sick from infected wounds or skin infections? The reality is that our skin is so effective as a defense perimeter, that usually pathogens are easily repelled here. Most of the infections that you will consciously deal with in your life will enter your body elsewhere, in another kingdom. A kingdom that has to solve one of the hardest conundrums of your entire defense network. And it is the place where your most dangerous enemies strike you.

Part 3

Hostile Takeover

24 The Swamp Kingdom of the Mucosa

Whatever you do in life, it is not possible to exist and function without the world and the things it offers. There is no pillow fort, no remote cabin in the woods, no teenager and their computer, no worldwide social distancing intense enough to protect you from the fact that you need to interact with the world. At the very least, you need a constant influx of food and so a minimal amount of interaction with the outside is unavoidable.

Your body faces the same problem because your cells need oxygen and nutrients to keep themselves going and discard dangerous waste that is a byproduct of the cellular metabolism. In other words, resources need to get from the outside to the inside while garbage needs to get from the inside to the outside. So your body can't be a closed system—it is unavoidable to have places where your insides directly interact with the outsides.

But these places are dangerous weak points that allow uninvited guests to sneak into the continent of flesh. And indeed, the vast majority of pathogens that make you sick enter where the interactions with the outside happen. In the long tube that begins in your mouth and ends in your butt, or in the many side tunnels that lead into the cave systems that make some sort of exchange possible.

As we mentioned at the beginning, your lungs, your guts, mouth, and respiratory and reproductive tracts are really just outsides wrapped up in the insides. These insides are lined with what you could call your "inside skin." Unfortunately the correct name is *Mucosa*. But to make it a bit more badass we will call it *the Swamp Kingdom of the Mucosa*.

The swamp kingdom needs to solve the immense problem of being easy to cross for nutrients and substances the body wants to get rid of, while being hard to cross for pathogens at the same time. This means that in and around the swamp kingdom the immune system has to be different than in the rest of the body.

While most of your continent of flesh is pretty sterile and devoid of microorganisms, *devoid of other*, your swamp kingdom is constantly in contact with all sorts of *other*—pieces of foodstuffs that need to be taken in, indigestible stuff that just passes through, friendly bacteria that get a free pass and are allowed to stay in your gut, all sort of particles that flow through the air and are breathed in, from pollution to dust.

And of course with all that come countless unwanted visitors that try to sneak in and pass the defenses. Some innocent travelers that just got lost, some of them dangerous pathogens that have specialized in hunting humans. This makes the job of the immune system around these places extra hard and the balancing act it has to perform even harder. Because in the Swamp Kingdom of the Mucosa, your immune system has to be somewhat tolerant.

In contrast, in most parts of your body, your immune system is not tolerant *at all*. Like when you cut yourself and bacteria invade your soft tissue, the immune system reacts with maximum violence and anger. A bacterium below your skin or flesh is unacceptable and needs to be killed immediately no matter the costs. Around your mucosa this is not possible. Just imagine your immune system attacking every little unwanted bacterium sitting on a piece of food with as much anger as the bacteria in our rusty nail scenario.

Imagine it reacting violently to every little flake of dust that you breathed in. No, the immune system of the swamp kingdom can't be as aggressive as the immune system in other parts of the body or it would simply destroy the places made for the exchange of gases and resources, which can make your life miserable or even kill you (as it indeed does for many people who suffer from autoimmune diseases or allergies, but more on that later). In the mucosa the Immune System had to learn to tread more lightly and to act as locally as possible when provoked. Yet at the same time, the mucosa is the most vulnerable area of the whole body, so the immune system also can't be incompetent or super chill here. A truly hard nut to crack.

So the first countermeasure to prevent invasion is to be a horrible and deadly place for unwanted microorganisms. So the mucosa employs a number of different defense systems.

If the skin is like a vast desert and a nearly impassable wall protecting the border of the continent of flesh, the mucosa is like a vast swampland, with deadly traps and groups of defenders patrolling. Easier to pass than the desert and border wall of the skin, but still not *easy* to pass. OK. So what is the mucosa and how does it defend you?

The first line of defense employed by the swamp kingdom is the swamp itself. The *mucus layer*. Mucus is a slippery and viscous substance that behaves a little bit like watery gel. You know it as the slimy stuff in your nose that becomes especially visible and disgusting when you have a cold—but it is actually all over your insides—in your gut, your lungs, your respiratory system, your mouth, the inside of your eyelids.

It covers all of your surfaces that interact with the outside that are wrapped inside you. Mucus is continuously produced by *Goblet Cells*, which are not important for the story of the immune system but they look really funny. Imagine them as weird squished worms that have to vomit all the time to create the mucus layer.

This slimy mucus serves multiple purposes—in the simplest sense it is just a physical barrier so intruders have a harder time reaching the cells that it covers. Imagine swimming in a pool filled up with slime. And then imagine trying to dive to the bottom, only the pool is 300 feet deep (please excuse this mental image). And mucus is not just a sticky barrier but also filled with unpleasant surprises similar to the desert kingdom: salts, weaponized enzymes that can dissolve the outsides of microbes, and special substances that sort of sponge up crucial nutrients that bacteria need to survive, so they starve to death inside the mucus.

In most places your mucus is also saturated with deadly IgA Antibodies. So just by itself, the slimy part of the swamp is not a very welcoming place. But it does not just protect you from outside intruders, it also protects your body from itself. For example, have you ever asked yourself how it is possible that you have a sack filled with literal acid inside of you? Well, the mucosa inside your stomach acts as a barrier that keeps the acid at a distance and protects the cells making up your stomach wall.

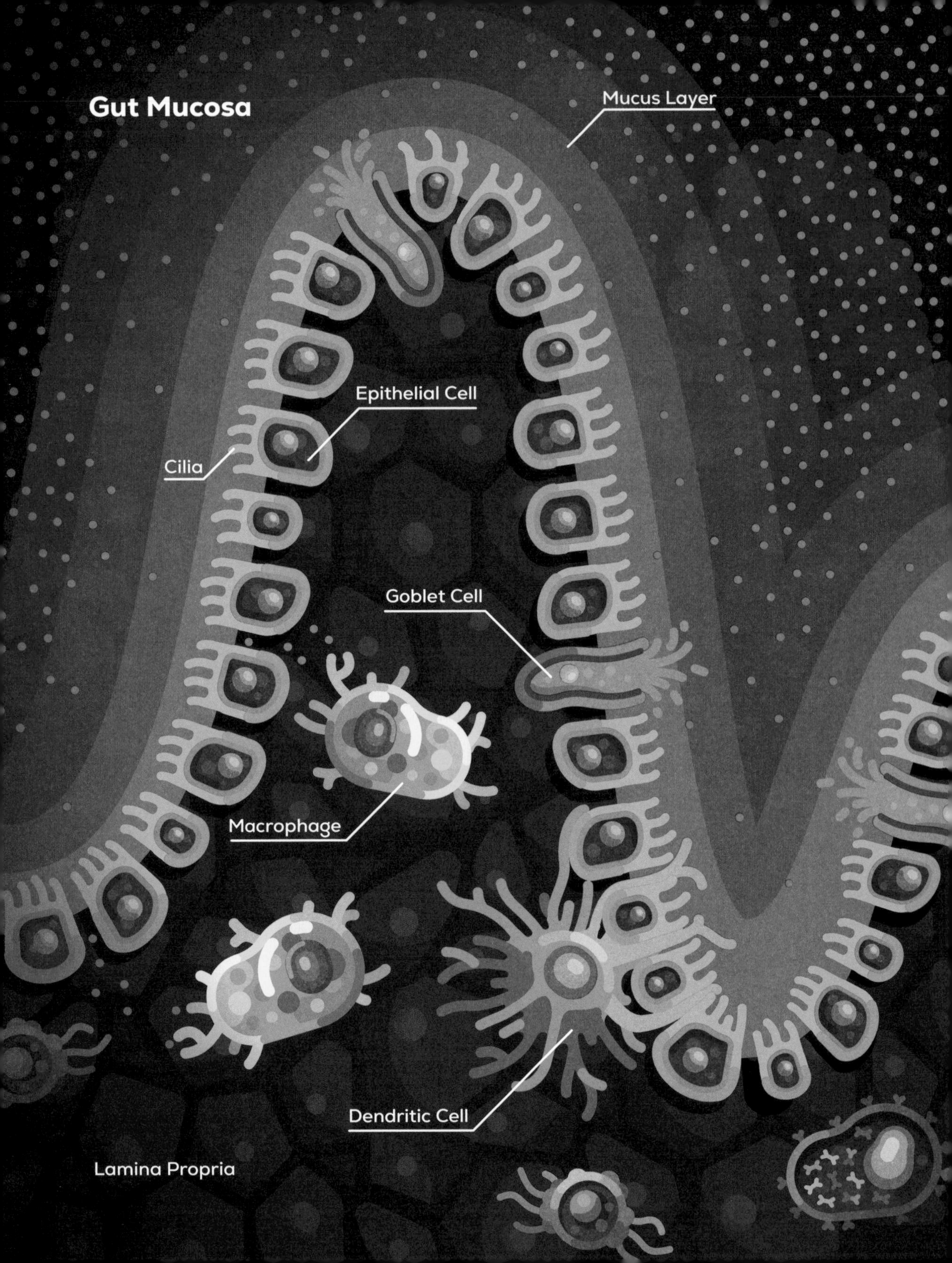

Gut Mucosa

Mucus Layer

Epithelial Cell

Cilia

Goblet Cell

Macrophage

Dendritic Cell

Lamina Propria

Commensal Bacteria

M-Cell

IgA

B-Cell

Relax...

Helper T-Cell

But the mucus is not just slime sitting there—it also moves. A vast network of fine *cilia*, tiny organelles that look a bit like hair, cover the membranes of the special cells that make up the first layer of the mucous membrane: **Epithelial Cells**. These cells are the equivalent of your skin cells, if you want. They are the cells that directly sit on the border of your mucous membranes, only covered by the slime. They are your "inside skin" cells.

In some places there is only a single layer, one cell thick, between the slime and the inside of your body. Epithelial Cells don't have the luxury that the skin has, where hundreds of cells are layered on top of each other. So Epithelial Cells are no pushovers. Although they are not technically cells of the immune system, they play a crucial role in your defense as they are especially great at activating your immune system and at calling for help with special cytokines. You can imagine them a bit like a citizen militia, no match for an enemy army but a very helpful addition to your defenses in case of an invasion.

And one of their jobs is to move the slime with the hairlike cilia that cover their membranes. Some microorganisms use cilia to move themselves around, while your epithelial cells use them to move the slime that covers them by sort of "beating" in unison. The direction depends on their location. In your respiratory tract, your nose, and lungs, the slime is moved either directly out of the body through your mouth or nose or via a slight detour it is swallowed and ends up in your stomach.

We swallow a fair amount of this slime in our lives and as disgusting as this may be, it is a pretty good system. After all, your stomach is an ocean of acid that the vast majority of pathogens are unable to survive. In your gut the direction should also be clear—things come from the stomach and move to your anus, where all things that enter your mouth must leave eventually.

But the Swamp Kingdom of your Mucosa is not really a single kingdom—it is much more like an alliance of different kingdoms that are all very different from each other but working together with a common goal. Which makes sense. The Desert Kingdom of the Skin may vary in thickness between the bottom of your foot and your lower back, but its job is more or less the same. In contrast, the mucosa in your lungs has a very different job from the mucosa in your gut, which has a completely different job from the mucosa in the female reproductive tract. And according to the different specializations of the realm, the immune system that protects it works differently. Before we move on to our next great enemy, the virus, let us take a look at the weird kingdom that is your gut and how it deals with the literal trillions of bacteria living there.

25 The Weird and Special Immune System of Your Gut

YOUR INTESTINES ARE A VERY SPECIAL PLACE FOR YOUR IMMUNE SYSTEM because a lot of complicated challenges need to be managed here to keep the body healthy and functioning.

Again imagine your guts as a long tube reaching through you, trapping a bit of outside inside you. On this outside, on your gut mucosa, around thirty to forty trillion individual bacteria from around 1,000 different species and tens of thousands of species of viruses make up your gut microbiota (the vast majority of viruses in your gut are hunting the bacteria living there and not interested in you).

There are a lot of things about the interactions and functions of your Immune System and the gut microbiome we don't understand yet. We know that many diseases and disorders are associated with these interactions being out of balance, but a lot of research needs to be done before we will be able to fully understand all these relationships. It is likely that the next few years will reveal a lot of exciting things.*

---

* This is the footnote about poop transplants and the time during World War II where German soldiers ate camel poop. So it is known that your gut microbiome and how healthy it is has a strong association with how healthy you are and what you can handle. So in the last few years the so-called poop transplant has become a thing in modern medicine. Which means what you think it means—poop from a healthy person, which carries a healthy dose of their gut microbiome, is administered via a special pill to a patient. (Or, if you have to know, via a long tube that drips down the poop from the back of your throat into your stomach.)

It is not entirely without risks but, for example, it is very effective in fighting *Clostridium difficile* infections, which is a nasty bacteria that are ubiquitous in nature and can also live in small numbers in your gut. In certain cases, like when a patient has to take large doses of antibiotics that kill a lot of the bacteria of your gut, it can take over and become a pathogen that can cause everything from diarrhea, vomiting, or in the worst cases life-threatening chronic inflammation of your gut. They are very resistant and sturdy bacteria and today many strains have become resistant to many

In this chapter we'll explore a little how it is possible that this coexistence with so many guests is even possible.

So first of all, the immune system of your intestines is a semi-closed system that tries not to mix too much with the immune system in the rest of the body. In a sense, it is a bit like Switzerland, which is surrounded by countries that are part of the European Union. Sure it is part of Europe, but still does its own thing to a certain degree and is technically independent. And in a way the swamp kingdom of your gut is a bit like that, because it needs to do a lot of things differently.

The greatest challenge it has to contend with is that the defense perimeters of your intestines are constantly breached. There is a never-ending stream of attacks and your gut immune system has to perpetually react and separate friends from foes, more than in any other place of your body. Because as you can probably imagine, your intestines are a busy place. Aside from the trillions of organisms that form your gut microbiome, there is all the stuff that you put inside your mouth.

Food begins its journey of becoming part of you and your cells by being

---

antibiotics, which can make it very hard to get rid of them. One of the things making it possible for *Clostridium difficile* to become a problem in the first place is a weakened natural gut microbiome. Poop transplants have been shown to have a high chance of restoring the natural balance and help the patients to get rid of the invaders on their own.

This is basically the idea behind poop transplants but it is not really a new idea. There is evidence that thousands of years ago eating the feces of animals was used to treat stomach and gut-related problems and diseases. Which brings us to World War II and the failed conquest of North Africa by the German army. Among problems like land mines and, well, losing, a huge problem the German troops faced was dysentery, a chronic inflammation that causes horrible cramps and dizziness, diarrhea, and dehydration (of all the places, the desert is not the place to lose a lot of water) and can be deadly.

The problem was simply that the soldiers were not used to some of the local microbes and since this was a time before antibiotics they had little recourse. A medical science unit that was sent to find a way to help the suffering men out discovered something peculiar though. Locals who got sick did not die of dysentery but instead collected the poop of camels and ate it. And to the utter astonishment of the observers, usually within a day, the sickness disappeared.

The locals had no idea why it worked, just that it did and that it had been done for generations. So the German doctors examined the camel poo and found *Bacillus subtilis*, a bacteria that suppressed other bacteria, among them the kinds that caused the dysentery. They cultured these bacteria in large amounts and administered them to the sick and dying troops, alleviating the problems of the German army somewhat. While this was a great moment for science, it did not stop the North Africa campaign from being a huge failure though.

ground into pieces by our teeth and being saturated and prepared by your saliva. Saliva contains a number of chemicals that help break down your food, so digestion really begins right after you begin eating. This makes sense, because as food is funneled through you, there is only a limited time window to extract the resources, so better start as soon as possible. After the ground-down foodstuffs are swallowed, they get to have a moment in an ocean of acid in your stomach. Which is not only helpful for your digestion and helps break down tough meat or fibrous plants, many microorganisms don't like being submerged in literal acid and die here, making the job of your immune system much, much easier.

After the stomach the journey continues through your intestine, which is between ten and twenty-three feet long and constitutes the longest stretch of our digestive tract. Over 90% of the nutrients you need to survive are absorbed here. And here a lot of the bacteria buddies you need to survive spend their time assisting with breaking down the food even more and enabling your body to take up its nutrients. But these are not just any bacteria. Millions of years ago your ancestors made a fragile deal with a team of microbial species—humans provide them with a long, warm tunnel to live in and a constant stream of stuff they can eat, and in turn they break down carbohydrates that we can't digest and produce certain vitamins that we can't make ourselves. The bacteria of the microbiome are tenants of sorts and these resources are the rent they have to pay.

These bacteria are called *commensal bacteria*, which comes from Latin and means something like "together at the same table." Like the barbarian bacteria hordes on the Desert Kingdom of the Skin, commensal bacteria and you are friends. The deal works best if they don't harm you and your immune system doesn't kill them. So to keep everything nice and peaceful, the bacteria of your gut live on top of the mucus layer of your gut, just like bacteria on your skin live on top of your skin. As long as your gut bacteria respect this separation and don't try to dive deeper and get to the epithelial cells, all is good. But of course things are not that easy.

Bacteria are not *actually* our friends and they don't know anything about any kind of deals and they don't respect anything. Because of the sheer vastness of our intestines and their incredible numbers, every second of your life, a bunch of commensal bacteria wander deeper into your body. And this is a problem, because if they were to enter the bloodstream and enter your actual *insides*, they could do horrible damage or even kill you. So the mucosa of your intestines is built in a way to prevent that.

In a nutshell there are three layers: First a layer of mucus filled with antibodies, defensins (we met these before on the skin, the tiny needles that can kill microorganisms), and other proteins that kill or damage bacteria. In the gut it has to be pretty thin and somewhat porous because all the nutrients from your food need to pass inside, and if the first protective layer were too good, you would starve to death.

Below the mucus layer, the intestinal epithelial cells are the actual barrier between the inside and the outside. Similarly to your lungs, the layer of epithelial cells protecting your insides is only ONE cell thick. To be a bit better at protecting your insides, the intestinal epithelial cells are extremely well-connected with each other. Special proteins glue them together firmly, so they can be as good a wall as possible. Your immune system monitors this area and is especially annoyed with any kind of microorganism that is trying to attach itself to the epithelial cells.

This is actually happening constantly, every second of your life—a bunch of commensal bacteria passing the defensive wall. So below the epithelial wall is the third layer of the gut mucosa, the *Lamina Propria*, which is the home of most of the immune system of your gut. In the Lamina Propria, directly below the surface, special Macrophages, B Cells, and Dendritic Cells are waiting to greet the unwelcome guests:

Your intestinal immune system really does not want to cause inflammation if not absolutely necessary, because inflammation means a lot of extra fluid in the intestines, which you experience as diarrhea. Diarrhea doesn't just mean watery poop, but also damage to the very sensitive and thin layer of the cells that take up the nutrients from your food. And diarrhea can rapidly dehydrate a patient to dangerous levels.

Unbeknown to most people, diarrhea is still a huge killer, responsible for about half a million dead children each year. So millions of years ago when we evolved, our bodies and immune systems learned to take inflammation in the intestines very seriously.

As a consequence the Macrophages guarding your intestines have two properties: Firstly, they are really good at swallowing bacteria. And secondly, they do not release the cytokines that call in Neutrophils and cause inflammation. They are more like silent killers, casually eating bacteria that cross the line but without making a fuss about it.

The Dendritic Cells of your gut behave in a special way too. Many of them sit directly below the layer of epithelial cells and squeeze their long arms between them, reaching right into the mucus of the gut. This way they can constantly sample the bacteria that are cheeky enough to not stay in their lane but venture too deep.

At this point lies a huge mystery of immunology that promises another Nobel Prize for the person or team who solve it one day: How do Dendritic Cells know if the bacteria they sample in the gut are dangerous pathogens or just harmless commensal bacteria? Well, right now we don't know but we know that when Dendritic Cells sample commensals, they order the local immune system to chill out and not be too annoyed by their antigens.

And there is more. Around your gut, special types of B Cells produce nothing but large amounts of IgA Antibodies, the Antibody that works especially well in mucus.

IgA Antibodies are specifically made for this kind of environment—for one they can be passed right through the barrier of epithelial cells and saturate the mucosa of the gut.

And IgA does *not* activate the complement system and does not trigger inflammation, which are both very important here. IgA is really good at something else though: with its four pincers that reach in opposite directions, it is an expert at grabbing two different bacteria and clumping them together.

So a lot of IgA can create huge clumps of helpless bacteria that are transported out of the body as part of your poop. All in all, around 30% of your poop consists of bacteria—and a lot of them have been clumped up by IgA Antibodies (most disturbingly, around 50% of them are still alive when they leave you). Your gut immune system quietly makes sure the visitors on your inside and outside are kept in check. So with these mechanisms and special cells, your immune system keeps the mucus free of overly ambitious friendly bacteria but also makes sure that it does not cause damage by overreacting. Your gut immune system really is a peacekeeping force.

All of these mechanisms are a horrible idea though if there are real invaders, like pathogenic bacteria that somehow could survive the brutal acid ocean of the stomach and reach the intestines intact. To catch these serious enemies as early as possible, your gut has a type of special lymph node called Peyer's patches that are directly integrated into your intestines. *Microfold Cells* (the same cells that we briefly met in your tonsils) reach directly into the intestines and take samples of things they think might be interesting for the immune system to take a look at. In a way, they are a sort of elevator cell that picks up passengers and transfers them directly into the Peyer's patch, where adaptive immune cells check out everything that goes on in your gut. This way your intestines have a superfast immune screening that constantly monitors the population of bacteria in your gut mucosa very closely.

OK, enough of bacteria and how they interact with your body. It is time to meet one of the most common invaders that you have to deal with in your life. An enemy that does not just invade the body but goes one step further and directly infects the cells themselves, where it can stay hidden from immune cells, to do its dirty work. This is such a smart and dangerous strategy that your immune system had to develop completely different strategies and weapons.

So let's explore your (arguably) most sinister enemy, the virus.

26 What Is a Virus?

VIRUSES ARE THE SIMPLEST OF ALL SELF-REPLICATING SORTS OF LIVING things, although, depending on who you ask, they may not even be considered alive. We talked about the lack of consciousness and awareness of your cells. That they are just really complex piles of biochemistry that do what the genetic code and the chemical reactions between their parts compel them to do. Bacteria are the same, protein robots able to do amazing things, although, in a sense, they could be considered a bit less sophisticated.

Viruses are not even that. The fact that a virus is able to do anything at all is equally depressing and fascinating. A virus is not much more than a hull filled with a few lines of genetic code and a few proteins. They completely rely on proper living things to stick around.

And they got extremely good at that.

It is not clear yet when or how exactly viruses came into existence, but it is very likely that they are ancient and already existed when the last common ancestor of all living things on earth was alive, billions of years ago. Some scientists think viruses were essential steps in the emergence of life, others think they are the result of an ancient bacterium taking the path of becoming simpler instead of more complex around 1.5 billion years ago. According to this idea, they were living beings that opted out of the life game and decided to save the effort and energy of constructing a functioning cell and instead started to rely on others to do all the hard work.

Whatever the truth is, viruses turned out to be incredibly successful. In fact, viruses are arguably the most successful entity on the planet. It is estimated that there are 10^{31} viruses on earth. Ten thousand billion, billion, billion individual viruses.[*]

---

[*] If we somehow collected them and laid them end to end, they would stretch for 100 million light-years—as many as 500 Milky Way galaxies put next to each other. In the oceans alone, every

Various Viruses

Viruses are arguably the most successful entities on planet Earth. They also look very funny.

 Spike Protein

 Capsid

 Lipid Envelope

 DNA/RNA

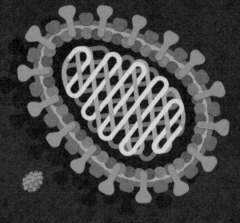

Influenza A Virus

Adeno Virus

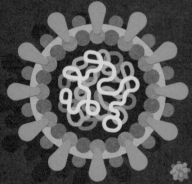

Corona Virus (SARS-CoV-2)

Ebola Virus

How did viruses become so successful, how did they do this? Well, in a sense, they don't do anything at all. They don't have a metabolism, they don't react to stimuli, and they can't multiply. Viruses are so basic that they have no way to actively do anything. They are literally particles floating around in the environment and have to rely on passively stumbling into victims by pure random chance.

If all other forms of life were to go extinct, viruses would disappear with them. So they need cells, proper living, active cells that do all that being-alive stuff for them. Some scientists even suggest that we consider a virus particle more as a reproductive stage, like a sperm cell, and a cell that is infected by the virus as its true living form. In any case, viruses are specialized to be vicious and sneaky intruders because obviously cells don't want to be infected by them. The main thing a virus needs to be able to do to thrive is to get inside cells. And for that they abuse a weak point of all cells that living things will never be able to completely protect against: They attack receptors.

We already talked a bunch about receptors, they are the protein-recognition parts that cover about half of the surface of cells. But receptors can do much more. They are used for interacting with the environment, to transport things from the inside to the outside and vice versa, and they are absolutely essential. The hulls of viruses are spiked with special proteins that can connect to a receptor type on their victims' surface. This means that viruses can't attach to just any cell—only to the ones that have a receptor they can attach to. In a sense every virus has a lot of puzzle-piece proteins that can only connect to a cell if it happens to have the correct puzzle-piece receptor.

Viruses are specialists, not generalists, and have preferred prey. Which is

---

single second, one hundred thousand billion, billion cells get infected by viruses. So many, actually, that up to 40% of all bacteria in the oceans are killed by virus infections every single day. And even more, even your most intimate self is not safe from viruses: About 8% of your DNA is made of remnants of viral DNA. We'll stop with the large numbers now because nobody can picture this stuff anyway. Let us just agree that there are a whole lot of viruses on earth and they seem to be doing quite all right. The fact that some apes with pants are discussing if they are alive or not could not be more irrelevant to them.

good because as we have established, there are a lot of viruses—but only about 200 different species infect us humans.

Once a virus gets in contact with the kind of cell it is looking for, it quietly takes it over. How a virus does this varies a lot from species to species, but in general a virus transfers its genetic material into its victim and forces the cell to stop making cell stuff. It is turned into a virus production machine. Some viruses keep their victims alive as sort of permanent living virus factories while others use up the cell as fast as possible. Usually for about 8 to 72 hours, the resources of the cell are turned into virus parts that get assembled into new viruses, until the cell is filled up, top to bottom, with hundreds to tens of thousands of new viruses.

Enveloped viruses leave the cell by budding from it, which means that they "pinch off" a bit of the cell's membrane and use it as an extra protective hull. Other viruses force the infected cell to dissolve and spill out its insides, including the new army of viruses it was brainwashed into building, which then go on to infect more cells.

If cells were conscious, viruses would be terrifying to them. Imagine spiders that don't crawl on walls, but passively float around the air, hoping to get into your mouth when you are not careful for a moment, crawling into your brain and forcing your insides to produce hundreds of new baby spiders until all your body is filled with them. And then your skin would burst open and all these new spiders would try to get your family and friends. This is literally what viruses do to cells.

Pathogenic viruses are excellent at circumventing the immune system because they have a superpower: Nothing multiplies as fast as they do. And that also means that nothing mutates or changes as fast as viruses. They are basically impossible to beat on that front because they are sloppy and careless. Viruses are so basic that they lack most of the intricate safeguards your cells have to prevent mutations, so they mutate *all the time.*

In general, the chance that a mutation is bad for an organism is higher than the chance that it is positive. But viruses don't care: Through the sheer incredible rate of reproduction and high numbers of individuals they produce in each reproductive cycle, with each infected cell, the chances that among a few thousand mutations, one is extremely beneficial and able to

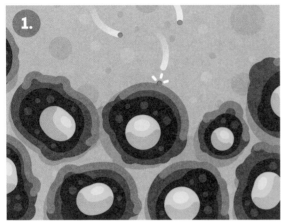

A virus succesfully connecting to Cell Membrane.

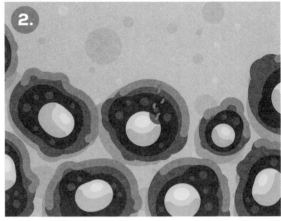

It successfully enters and takes over the cell.

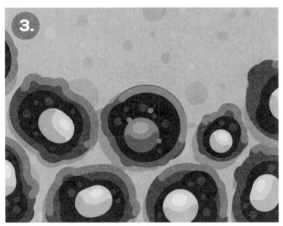

The virus uses the resources of the cell to make more viruses.

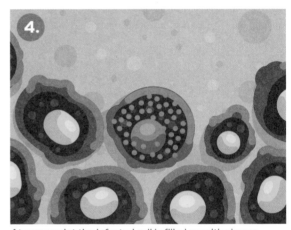

At some point the infected cell is filled up with viruses.

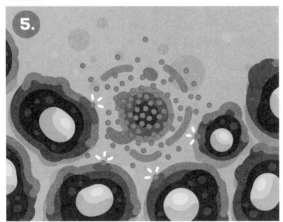

The cell dies and bursts open, releasing dozens to hundreds of new viruses.

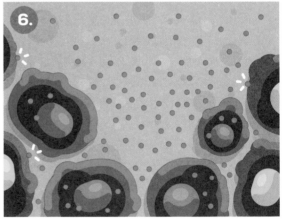

Neighboring cells are taken over, the cycle repeats.

make a virus significantly better suited to survive is pretty high. It's the old evolution, brute-force, throw-shit-at-the-wall-until-something-sticks approach. And it's quite effective.*

Your immune system can't rely on the same weapons to fight a viral infection that it uses to fight bacteria as both the enemy and its tactics are very different. A virus is smaller and somewhat harder to detect than bacteria because it doesn't have a metabolism that releases garbage chemicals that can be picked up by immune cells. And it hides inside cells for most of its life cycle and tries to manipulate infected cells to trick the immune system to stand down. It can change much more quickly than bacteria and a single virus can turn into ten thousand within a day, turning on exponential growth rapidly. Pathogenic viruses are terrifyingly dangerous enemies.

So it is no wonder that your immune system has invested heavily in anti-virus defenses.

But before we get to know our weapons, let us visit another mucosa kingdom, the main entry point for viruses. The majority of pathogenic viruses enter your body via your respiratory mucosa. And this makes sense—as we briefly talked about, your Desert Kingdom of the Skin is a really, really bad place to be if you are a virus that wants to invade human cells. Layers and layers of dead cells stacked on top of each other. In contrast, the mucosa of your lung is a very inviting entry point for a virus. This does not mean it is easy to enter here—just like the skin, the body created a powerful defensive kingdom here.

---

* Actually this is the only trick evolution has. It tries a lot of things and whatever does not die before it makes a few offspring gets another attempt at making offspring before dying. Repeat this often enough and you get the amazing variety of living things on earth. And new strains of cold viruses every season. So basically it's a mixed bag.

27 The Immune System of Your Lungs

THOUGH IT'S FUN TO IMAGINE, YOUR LUNGS ARE ACTUALLY NOT BIG BALloons, but in a way much like dense sponges with countless nooks and crannies. The parts of your lungs that do the actual breathing have an enormous surface area, in excess of 145 square yards (120 square meters)—more than sixty times the surface of your skin.

This vast space constantly interacts with the environment as you inhale a couple thousand gallons of air every day. As a consequence, your lungs are one of the most exposed places of the whole body. Each breath takes in about a pint (500 milliliters) of air, made up not only of the oxygen you need but also a few other gases that your body doesn't care about, and a plethora of particles. Exactly what stuff you are breathing in, and how much of that stuff, depends strongly on where you are in the world.

While in the cold of the Antarctic the air will be as fresh as it can be, consisting mostly of clean atmosphere. Walking on the busy streets of an inner city, you breathe in a wild mix of toxic exhaust gases, all sorts of particles from cars, and other aggressive materials like asbestos or the rubber abrasions from tires. Aside from this artificial pollution, air can carry a large number of allergens like the pollen from various plants or the dust in our homes, spiked with the droppings of mites.

Bacteria, viruses, and the spores of fungi are also riding on these particles or fine droplets of water, or are just floating around by themselves, looking for a new home. So the cells lining your lungs are constantly confronted with an onslaught of toxic chemicals, particles, and microorganisms. While in other areas of the body the immune system would react massively if it were confronted by this explosive mixture, damaging tissue without that

Respiratory System

The defenses of your breathing apparatus are a balanced system. It is able to fend off intruders and clear up the pollution while still allowing for the exchange of gases.

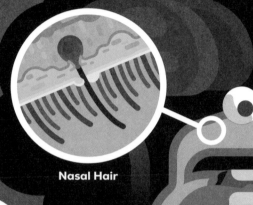

Nasal Hair

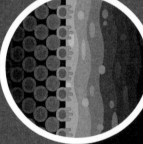

Mucus Layer

Epithelial Cells

Alveolar Macrophage

much regard, in the lungs this is not a great option. No matter what you do, you can't stop breathing.

So your immune system has to be more careful here, less brutal. A balanced system had to evolve in your lungs—able to fend off intruders and clear up the pollution while still allowing for the exchange of gases.

The defenses of your respiratory system begin in the nose, with a pretty large filter of actual hair—not useful against anything small but meant to keep big things from entering. Like large enough dust particles or pollen for example. Then, as in any mucosal environment, the mucus is covering the surfaces and in your respiratory system can be rapidly expelled by the explosive sneezing reflex.

The mucus is being moved constantly either outside or swallowed. In the deeper parts of your lungs these mechanisms are not useful though because to breathe, your alveoli, tiny sacs full of air, can't be covered by mucosa or breathing would not be possible. So at your deepest and most vulnerable places in your lungs, there is literally only a single layer of epithelial cells between the inside and the outside and nothing else. Talk about an exposed area. A perfect target for all sorts of pathogens.

To keep the area safe a very special type of Macrophage is stationed here: The *Alveolar Macrophage.* Its main job is patrolling the surface of your lungs and picking up trash. Most detritus and other unpleasant stuff are caught in the mucosa of the upper respiratory system, but some of it still reaches the deeper parts. Alveolar Macrophages are extra-chill Macrophages. They are much harder to provoke and activate than their cousins in your skin. In airways they downregulate other immune cells like Neutrophils and make them less aggressive. But most importantly, they tone down any sort of inflammation. Because the thing you really don't want in your lungs is fluid.

There is evidence that your lungs might have a microbiome (which means a collective of microbes that live in your lungs), or at least some sort of transient community of organisms that live in your lungs and is tolerated. But in contrast to the gut microbiome, we still know very little about the lung microbiome. This has multiple reasons. For one, on the microscale breathing is a hurricane-level storm that occurs constantly, so it is much harder for microbes to make their home here than in the chill gut. Then there are much fewer free resources and friendly bacteria have a much harder time making a living. But one of the biggest problems holding us back is that it is pretty hard to collect samples from the deep lung microbiome. You really need to appreciate how easy it is to collect stuff from the gut: Your gut is a long, wide tube and every day a happy sample of everything inside leaves through your butt. Your lungs are not nearly as cooperative and it is also pretty hard to collect samples from the deeper parts without contaminating them on the way out. So there is still a lot to learn about the microbiome and its interactions with the lung.

What we know for sure, though, is that a lot of the most common and dangerous pathogenic viruses that infect humans use the respiratory system as an entry point. So now that we have gotten an idea of the environment of your lungs, let us see what happens if they are infected and learn what sort of special defenses your immune system has found to wipe them out.

28 The Flu—The "Harmless" Virus You Don't Respect Enough

"ONLY THREE MORE DAYS UNTIL THE WEEKEND!" YOU THINK, AS YOU ENTER the break room where one of your colleagues is making coffee. Just as you pass her, she suddenly coughs violently, quickly covering her face with the crook of the arm, but not fast enough—the first cough hit the air unhindered and a fine cloud, made up of hundreds of droplets, shot through the air. On the scale of cells, these droplets are not like bullets but more like intercontinental ballistic missiles, traveling a distance equivalent to continents in seconds. And they are not filled with nuclear warheads, but an equally dangerous load: millions of *influenza A* viruses that cause a disease we know as *the flu.**

The larger and heavier droplet warheads don't get very far, soon hitting the ground. But the lighter ones spread out through the air carried by favorable air flows. You don't notice any of this as you walk right through the cloud of droplets. You breathe in, and a few dozen of the virus-filled missiles are sucked into your airways and violently splash onto your mucous membranes, where they release their viral load. You just make your coffee without realizing what a serious sequence of events has just been triggered. A bit later, as you begin to consider getting another cup, the first virus takes over one of your cells.

It will be the first of billions.

The influenza A virus that you breathed in so casually belongs to one of

---

* The name influenza means "influence" in Italian and stems from a time in the Middle Ages when people thought that the influence of astronomical events could affect your health and cause diseases. For example, liquid flowing off the stars and into Earth and then humans somehow. Almost as crazy as the idea that the positions of stars when you were born have an influence on your character and personality traits.

The Cough

Hundreds of droplets, filled with millions of viruses, shoot through the air. The larger droplets hit the ground soon, but the lighter ones spread out through the air forming long lasting clouds, for clueless bystanders to breathe in.

Aerosol with viruses

the most powerful and consistently dangerous strains of the very annoying family of *Orthomyxoviridae*. Influenza A has specialized in infecting the epithelial cells of the respiratory system in mammals. Since this includes humans, influenza A has been responsible for four major influenza pandemics in the twentieth century alone, the most famous one being the Spanish flu that killed at least 40 million people. Lucky for you, the strain you just breathed in is not that deadly. On average the "regular" flu that we have gotten used to "only" kills up to half a million people each year.*

For the viruses that entered your respiratory system in the break room, a crucial timer begins. They have only a few hours to reach their goal because the environment of the swamp kingdom is slowly but surely destroying them. Various proteins or antibodies that float around here can dismantle them or render them useless and they are being carried away with the mucus layer that is constantly replenished. And so, many of the virus particles that you breathed in never reach their goal because they are caught and destroyed in time. But in truly dramatic fashion, a single one of the viruses reaches the cells below the protective mucus.

Your epithelial cells, the "skin" of your insides, have receptors on their surfaces that the influenza A viruses can connect to and manipulate to enter the cells. It takes the virus only about an hour to gain control over the cell by conquering its natural processes. Without knowing what it is doing, the cell carefully wraps the virus inside a package and pulls it deep inside towards its nucleus, the brain of the cell. Natural processes, again triggered by the cell itself, signal to the virus when it has reached its destination and when it has to release its genetic code and a bunch of different hostile virus proteins.

Within ten minutes the influenza tricks the cell into delivering its genetic material directly into the brain of the cell, the nucleus. Viral proteins begin to dismantle the cell's internal antivirus defenses, and with that, the cell has been conquered.

The influenza A virus is trying to directly take over the nucleus, which is

---

* The Spanish flu was special because it turned the tables a little bit—usually the flu kills mostly small kids and older adults but in this case the opposite happened. If you were a healthy adult in your prime you had the highest chance of dying from the Spanish flu. The disease was the hardest on the healthiest people because it made their immune systems flip out and lose all constraint, leading to an overall mortality rate of roughly 10%.

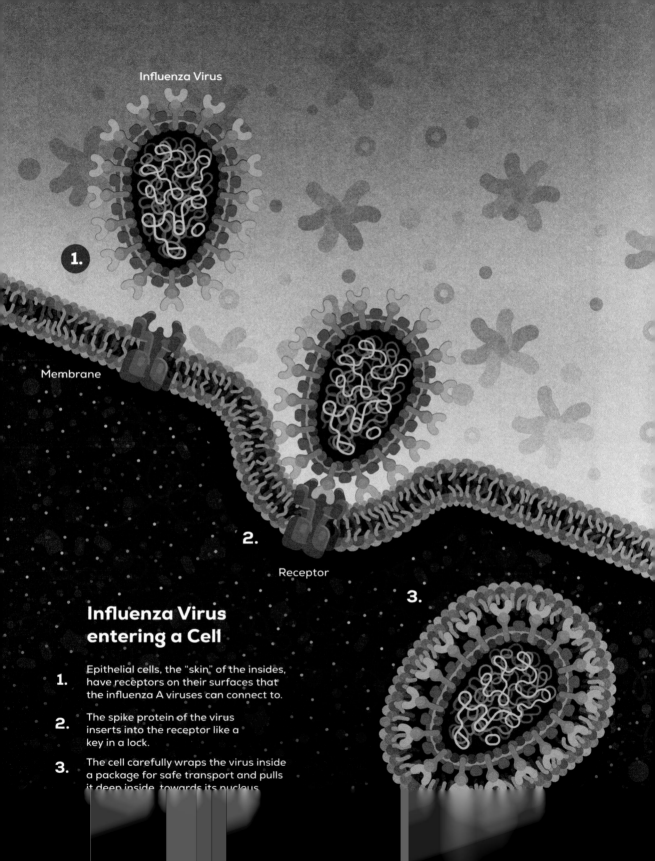

Influenza Virus

Membrane

Receptor

1.

2.

3.

Influenza Virus entering a Cell

1. Epithelial cells, the "skin" of the insides, have receptors on their surfaces that the influenza A viruses can connect to.

2. The spike protein of the virus inserts into the receptor like a key in a lock.

3. The cell carefully wraps the virus inside a package for safe transport and pulls it deep inside, towards its nucleus.

the brain of your cell, if you will. It stores the DNA, which carries the instruction manuals for all the proteins of the cell, but not just blueprints but also their production cycles. These proteins determine the development, function, growth, behavior, and reproduction of your cell. So whoever controls the protein production controls the cell itself. How does this work? Well, your DNA consists of smaller sections, your genes, and each gene is the instruction for one protein. To turn the instructions from a gene into an actual protein, this information needs to be transmitted to the protein production machinery in the cell.

How do genes transmit information? Well, they technically don't do anything because genes are just sections of your DNA. To communicate the information stored in a gene to the rest of the cell, living things use *RNA*. RNA is a complex and fascinating molecule that fulfills a variety of different and crucial jobs. The one we care about in this context is to act as messengers that relay the building instructions from genes to the protein factories of your cells.

And here viruses come in and screw everything up. Viruses try to take over this beautiful natural process in a multitude of ways, depending on their modus operandi. The influenza virus A, for example, just dumps a number of RNA molecules into the nucleus, where it pretends to be commissioned from your own genes and tricks the cell into building specific viral proteins. But of course the viral proteins are harmful and interrupt the production of healthy cell proteins and instead produce virus proteins, or in other words, virus parts.*

In our story, the influenza A virus that infected the epithelial cell has been successful and the fate of the poor cell is now sealed. It has become a

---

* With viruses we have now truly entered the intimate and mind-bending world of biochemistry. Cells are made from millions of parts moved by thousands of processes going on at the same time, in a complex and wonderful dance that we call life. Viruses interfere here in ways that are stunningly complex and complicated. If we were to go into detail, we would encounter viral proteins and molecules with horrible names like vRNPs, viral polymerase complexes like PB1, PB2, or PA, viral membrane proteins HA, NA, or M2, polypeptides like HA1 and HA2. This stuff is fascinating but it also requires a multiple-page-long discourse about the detailed inner workings of your cells and how viral parts interact and manipulate them. It is just a layer of complexity that is not necessary to get the principles at play here. You really need to remember only one thing: The virus is basically performing a hostile takeover of the machinery of your cell.

dangerous time bomb for your body, a protein robot that does not serve you any longer, but now serves a new and sinister master.

For a few hours, processes and production lines are changed and molded for their new purpose before mass production of new viruses begins. According to some estimates, a single cell infected by influenza A, is, on average, able to produce enough viruses to successfully infect twenty-two new cells before the first victim cell dies from exhaustion after a few hours.

If we assume this process plays out without resistance (and each virus infects only uninfected cells) then one infected cell becomes 22, which then become 484 infected cells. Then 484 turn into 10,648, which turn into 234,256, which turn into 5,153,632. In just five reproductive cycles, each taking about half a day, a single virus has turned into millions. (In practice this will not always pan out like that since your body will not just let this happen—but then again, you probably will have more than one influenza virus being successful in infecting your cells initially. So millions of infected cells is probably not far off.)*

Viruses are just something else when it comes to exponential growth. They play in their own league where they can multiply explosively in large bursts, rather than in a binary fashion like bacteria, for example.

Speaking of bacteria, in contrast to the very straightforward battlefield when we stepped on a nail, the situation is very different with viruses.

If you cut yourself and bacteria infect your wound, things are pretty straightforward: There is damage that immediately causes inflammation and attracts the immune system and there are a lot of enemies that don't exactly act super discreet but more like a bunch of drunk toddlers in a candy shop.†

---

* So as we are talking about millions of infected body cells, what does it mean to you in a practical sense? How much of your lung is infected at this point? How big is a patch of a million infected epithelial cells? Very roughly, a million infected epithelial cells have a surface area of half an inch (1.2 centimeters). This is roughly less than half of the surface area of a Lincoln penny or a Euro cent. In total your lungs have a surface area of about 84 square yards (70 square meters), which is just short of a badminton court. So actually only a tiny amount of your lungs has been infected at this point. It becomes scary again if you remember how tiny a cell is and how quickly this scaled up from basically nothing. If the virus was allowed to grow at this rate your whole lung would be infected in no time and you would be very dead.

† OK, OK, that is slightly unfair. Not all bacteria are like clumsy idiots, many pathogenic ones have arguably genius strategies for hiding and striking hard when the time is right. A very cool

Viruses don't want attention. The beginnings of an influenza A infection are less of a full frontal attack and more like an invasion by a bunch of commandos trying to stay undetected and silently taking out your defenses.

Think about the legend about the ancient Greeks trying to conquer the city of Troy thousands of years ago, the one with the wooden horse. If you imagine Troy as your body, the siege and attacks with open field battles in front of the city gates is what most bacteria do—running around screaming and having their heads bashed in by defenders who are very annoyed with them.

The influenza virus is more like the soldiers hiding inside the Trojan horse, trying to enter the city as sneakily as possible, doing everything to stay hidden. Once they are inside they wait until nightfall and try to sneak from house to house to kill the Trojan citizens in their sleep before they can alert the city guard about the invasion. Every house they take over becomes a base for the intruders that creates more invading soldiers, and every night more of them try to silently take over more houses and to kill more citizens in their sleep. OK, at this point the metaphor breaks down a little, but you get the gist.

In a nutshell this is a major feature of an infection by pathogenic viruses. This sneaky approach also means that the battlefield is very different in the case of a serious viral infection than what we encountered with bacteria. If we look around at the site of freshly infected lung tissue, we don't see anything. Just seemingly healthy cells doing their thing while hidden enemies are slitting throats and disabling defenses inside the cells. In a very real way, this makes viral infections much crueler and more insidious than bacteria barging into an open wound.

Pathogenic viruses are truly scary enemies. They attack your weakest

---

example is quorum sensing. In a nutshell, this means that pathogenic bacteria invade a tissue but are very discreet. Like they control themselves and the metabolism strictly while they divide, downregulating all sorts of metabolic products (bacteria poop) and hiding their dangerous weapons that could reveal them to the immune system. They do that by waiting for a chemical signal that tells them to attack at the right moment. When a critical mass is reached they suddenly and all at once stop their secretive behavior. Now they are no longer a small threat that could be taken care of easily but a formidable army, and they lose the constraint all at once. If they behaved like that from the get-go they would have been attacked and probably killed right away. So yeah, quorum sensing is pretty cool and bacteria have more than one strategy.

links and hide inside civilians, where they proliferate much, much more explosively than other pathogens and can infect countless cells with every new reproductive cycle. At the height of a virus infection, you can have billions of viruses inside your body. All of these special properties require your immune system to defend against them differently than against most bacteria.

But don't get too scared. Your immune system has evolved special antivirus defenses.

At this point, maybe a few dozen cells have been infected, but the first countermeasures are already booting up. This early on in the infection there is a struggle between your infected cells that want to alert the immune system and the virus that tries to silence them.

To return to our Troy metaphor, the peaceful slumbering citizens of the city wake up to the sound of enemy soldiers sneaking into their homes and trying to slit their throats. So before this can happen they run to the windows and try to alert the town guard by crying out a loud warning. But just as the citizens want to scream, the intruders violently pull them from the windows and silence them forever with stabs and cuts. A desperate struggle for control for every single house and every single citizen plays out. If the citizens win and manage to call for the guards, the immune system will awaken; if the sneaky intruders win, they will gain the time they need to create more warriors and become a real danger to the whole city.

OK, houses and soldiers and citizens, screaming and struggling and stabbing. What is actually going on here and what sort of thing is the metaphor describing? Once again we are about to encounter a wonderfully elegant solution to an incredibly complicated problem.

The first real defense of your body against viruses is *Chemical Warfare*!

29 Chemical Warfare: Interferons, Interfere!

JUST LIKE THE CITIZENS OF TROY DESPERATELY FOUGHT THE GREEK soldiers that snuck into their city, your cells fight tooth and nail against the influenza virus that is inside them.

The first step in this struggle for your cells is to be able to realize that
they have been invaded. And since the epithelial cells that line your mucous
membranes are such a prime target for virus invasions, they are actually
prepared! As we mentioned before, epithelial cells are a sort of militia. And
as such, they do possess pattern-recognition receptors similar to the toll-like
receptors that we got to know early in this book. These are the receptors of
the innate immune cells that can recognize the most common shapes from
enemies like viruses. Your epithelial cells have a bunch of different receptors
that scan their own *insides* for red flags.

If they connect to certain viral proteins or molecules they know that
something has invaded them and that something is massively wrong, which
triggers an immediate emergency response.

Right then, your body has to confront a serious problem with viral infections. The innate immune system is not nearly as effective against pathogenic
viruses as it is against bacteria. So in the case of pathogenic virus infections
(or bacteria who hide inside cells), your body desperately needs help from
the adaptive immune system to stand a chance of clearing an invasion.

But as we have learned by now, the adaptive immune system is slow and
needs a few days to wake up, which is not ideal if you consider how quickly
viruses multiply. So in case of a serious viral infection, your innate immune
system and the infected civilian cells need to fight for the most valuable
thing in the universe: Time. They need to slow down the infection and make
it as hard as possible for the virus to spread further to more civilians.

And now we finally get to the way your cells do this: chemical warfare.

We talked about cytokines a bunch in this book—the amazing proteins that transmit information, activate cells, lead cells to the site of a skirmish, or make immune cells change their behavior. In short, cytokines are the molecules that activate and guide your immune system. They also do this in the case of a virus infection, but they actually play an arguably greater role here.

If one of your cells realizes that it is infected by a virus, it immediately releases a number of different emergency cytokines to the cells surrounding it and to the immune system. Those cytokines are the civilian shriek upon seeing the intruders at the foot of the bed.

There are a lot of different cytokines that are released in this situation and they do a lot of different things but here we want to highlight a very special class: *Interferons*. Interferons got their name from "interfere." They are cytokines that are *interfering* with viruses.

In a sense you can imagine interferons as a warning echoing down the streets of the city, calling citizens to lock their doors and move furniture in front of it, to board up their windows and anticipate an attack by soldiers. Interferons are the ultimate *"get ready for a virus"* signal.

So when cells pick up interferon molecules, it triggers different pathways that make them change their behavior drastically. One important thing to understand here is that at this point it is impossible for your body to deduct how many viruses are present, how many cells they have invaded, or how many cells are already producing new viruses in secret.

So one of the first changes is for the cells to temporarily shut down protein production. Every moment of your life your cells are recycling and reconstructing their internal building blocks and materials to make sure every protein is in good shape and works as intended. So some interferons tell cells to chill out a bit and to slow down the production of new proteins. If a cell doesn't build a lot of proteins, it can't build a lot of virus proteins if it happens to be infected already. So basically just by ordering cells to slow down, interferon slows down the production of viruses considerably.

There are more examples of targeted interventions and we could go into even more detail here, as there are dozens of different interferons that do dozens of different things, but in the end it doesn't matter that much. What

Interferons

Epithelial cells recognize that they're infected with the help of receptors on their insides. To warn other cells and to win time they release special cytokines called "Interferons." When cells recognize these interferons, they shut down protein production to slow down the infection.

Infected Cell

Interferons

Plasmacytoid
Dendritic Cell

is important for you to take away here is that interferons interfere with every step of viral replication.

Interferons will seldomly eradicate an infection by themselves, but they don't have to. All they have to do is slow down the multiplication of new viruses by making cells in the vicinity much more resistant to the virus infection. And sometimes this response is enough to prevent the spread of a virus infection so efficiently that it goes nowhere and you'll never even know anything happened at all.

In the case of our break room influenza A infection, this is not the case unfortunately. The influenza virus has adapted to the human immune system and comes prepared. When it unloaded its genetic information to take over the cell, it also came prepacked with a bunch of different viral "attack" proteins. These weapons are able to destroy and block the internal defense mechanism of the infected cells. You can imagine these attack proteins as the daggers of the soldiers invading the houses—effective tools to prevent screaming (cytokine release) by doing a bit of stabbing.

So while the influenza A virus is not always successful in preventing the release of interferons, it is very good at delaying it and buying more time for itself. Isn't this fascinating if you think about it? Two very different enemies, a virus and a human cell, and both are struggling with each other for time.

Influenza A is very good at this struggle and so often a few dozen viruses become tens of thousands in hours. Still, the initial tactic of staying as hidden as possible has the downside that even if it is successful initially, it will fail after a while. It cannot stay hidden forever. The more cells the virus infects, the more civilians will be able to activate chemical warfare, the more civilian cells will die eventually, which triggers inflammation and activates the immune system by itself, and the more virus particles will float around in the fluids between the cells, triggering red flags. So even the sneakiest virus will be detected sooner or later.

Usually sooner, because chemical warfare triggers the next step on the escalation ladder of the antivirus section of your Innate Immune System: *Plasmacytoid Dendritic Cells.**

---

* Plasmacytoid Dendritic Cells have one of these horrible names in immunology that are not helpful at all. A thing about the immune system is that there are a lot of different subclasses of

These special cells spend their lives moving through your blood or camp out in the lymphatic network, scanning specifically for signs of viruses—panic interferons from civilian cells or just straight up viruses that float around in your fluids. In any case, if they do pick up signs of a viral infection, they activate and turn into chemical power plants that ooze out extreme amounts of interferons, alerting not only civilians to turn on their antiviral modes (shut down protein production, etc.) but also the immune system to activate and get ready for a proper fight. You can imagine these cells as a sort of traveling smoke detector: A pathogenic virus like influenza A might be able to suppress the natural chemical warfare response of its victims and stay under the radar. But Plasmacytoid Dendritic Cells are able to detect even subtle signs of their presence and amplify them considerably to sound the alarm.

Indeed they are so sensitive to the signs of a virus infection, that only a few hours after the first of your civilian cells have been infected, they have opened the interferon floodgates. This is so fast that a spike of interferons in your blood is usually the earliest sign of a virus infection, long before any real symptom or the virus itself is detectable. In our break room story, this has happened a few hours after the cough that infected you. On your giant human level, you have not noticed or thought about any of this, much less felt any symptom yet.

But while this is great and the onslaught of interferons begins to awaken the rest of the immune system, influenza A continues to spread rapidly throughout your respiratory system. Hundreds of thousands of viruses emerge, leaving first thousands, then millions, of dead and infected epithelial cells in their wake. At this point, the stealth approach is no longer necessary, the virus has already succeeded in buying enough time to replicate

---

cells. So there are a bunch of different Dendritic Cells, a bunch of different macrophages, etc. The thing is: This doesn't really matter. It would be much better if the Plasmacytoid Dendritic Cell would be called: "Chemical Warfare Cell." Or "Antivirus Alarm Cell." Or anything other than what it is actually called, because all of these names would describe it better. We will handle this by never mentioning this cell again after we explain it here because on the one hand, it is too cool not to mention that you have this special antiviral chemical warfare cell but on the other, it is confusing that there are special "Dendritic Cells" with completely different jobs than the regular Dendritic Cells we got to learn so much about. So once we are done talking about it here, we can all just die happily without learning any more details about it.

prodigiously. To invoke our Troy story one last time—the invading forces are spilling into the open in daylight. Soldiers, guards, and civilians are fighting in the streets. Your immune system has to fare better than the citizens of Troy, though, or the virus will quickly overwhelm your body.

Meanwhile the weekend has begun and you peel out of bed, ready to play videogames and do other very important things. But you notice that something is off: Your throat hurts and your nose is runny, you have a bit of a headache and a cough. Usually you feel hungry right away after waking up but today you don't feel like breakfast at all.*

You have a cold, you self-diagnose with unwarranted confidence.

"Just in time for the weekend, life is so incredibly unfair! Nobody has ever had it as hard as me and nobody ever will have it as hard in the future," you lament to yourself, expecting sympathy from the universe but receiving none. You rally. This is but a scratch! You will just pop a few aspirin and enjoy your free time, a cold will not stop you. You are right of course—a cold will not stop you. But this is not a cold.

While you are gravely mistaken about the nature of what is going on inside your body, the influenza A virus is rapidly gaining ground, spreading throughout your lungs. It now has become a proper infection that is dangerous and that is still not contained. Your immune system is already in full response mode, as you will soon notice. We alluded to the fact a few times already, your immune system is often the part that causes the largest amount of damage in an infection and influenza is no different. All the unpleasant things you are about to experience are the result of the desperate attempts to halt the brutal invasion of your lungs.

The battlefield, which now stretches from your upper to your lower respi-

---

* Why do you lose your appetite when you are sick? Well, you can blame the onslaught of cytokines that your immune system releases for that. The cytokines signal your brain that a serious defense is happening at that very moment that the body needs to conserve energy for. Because as you may imagine, mobilizing millions or billions of cells for a fight is a pretty resource-heavy operation. Digesting food actually requires a lot of energy too, so shutting these processes down is freeing up your system to focus on the defense. It also reduces the availability of certain nutrients in your blood that your invaders would love to get their tiny pathogenic hands on. This does not mean that you should actively try to starve out a disease. Not digesting is a short-term strategy not a long term solution and in chronically ill people, the lack of appetite can lead to dangerous weight loss. So if you feel hungry again you can eat something to refill your energy storage.

ratory tract, has become busy. Local macrophages clean up dead epithelial cells and swallow free-floating viruses if they stumble over them, while they release cytokines to call backup and cause more inflammation.

Neutrophils join the fight too, although their presence is a mixed bag (and there is still active research and debate among immunologists if they are actually helpful in case of virus infections or are doing unnecessary damage). Neutrophils seem to not really be able to fight viruses well, so their help is mostly passive: Being the unhinged warriors they are, they increase the level of inflammation.

Here the general role of the Innate Immune System to provide context and make overarching decisions becomes apparent again: Your soldier cells realize that they are dealing with a virus infection and that they need help on a larger scale, so they release another set of cytokines: *Pyrogens.*

Pyrogen loosely translated means "the creator of heat," an extremely fitting name in this case. Simply put, pyrogens are chemicals that cause *fever.* Fever is a systemic, body wide response that creates an environment that is unpleasant for pathogens and enables your immune cells to fight harder. It also is a strong motivator to lie down and rest, to save energy, and to give your own body and immune system the time they need to heal or to fight the infection.*

Pyrogens work in quite a cool way, in the sense that they directly affect your brain and make it do things. You probably have heard about the blood-brain barrier, an ingenious contraption that stops most cells and substances (and pathogens of course) from entering the very delicate tissues of your brain, to keep it safe from damage and disturbance. But there are regions of your brain where this barrier is partially penetrable by pyrogens. If they enter and interact with your brain they trigger a complex chain of events that basically cranks up the temperature by changing the internal thermostat of your body.

Your brain cranks up the heat in two main ways: For one, it may generate

---

* A lot of different substances can be pyrogens, from certain interferons, to special molecules released by activated macrophages to the cell walls of bacteria. But in the end, you just need to remember one thing: Your innate immune cells release substances called pyrogens that order your brain to make your body hotter!

more heat by inducing shivering, which is just your muscles contracting really quickly, which generates heat as a byproduct. And by making it harder for this heat to escape by contracting the blood vessels close to the surface of your body, which reduces the heat that can escape through your skin. This is also the reason why you can feel so cold when you have a fever—your skin is actually colder because your body is trying to really heat up your core and create unpleasant temperatures at the battlefield to make pathogens really unhappy.

Still, fever is a serious investment for your body as it costs a lot of energy to heat up the whole system by a few degrees, depending on how harsh your fever is. On average your metabolic rate increases by about 10% for every two degrees Fahrenheit your body temperature rises, which means that you burn more calories just to stay alive. While this might actually not sound all that bad to you if you want to lose a bit of weight, in the wild burning extra calories is not a great idea most of the time, it is an investment that an organism has to hope will pay off in the end. And it seems to pay off most of the time!

Most of the pathogens that like humans operate really well at our regular body temperature and the higher temperatures during fever make their lives much harder. Just imagine the difference between going for a run on a fresh spring morning in contrast to going for a run in the summer heat at noon without any shade. It is just that much more draining to do anything if you are too hot. So the increased body heat actually directly slows down the reproduction of viruses and bacteria and makes them more susceptible to your immune defenses.*

---

* OK, let us talk about one of the weirdest Nobel Prizes for medicine and how upsetting the past was and how great the present is. Syphilis is a sexually transmitted disease caused by spirochetes bacteria. Its possible symptoms are horrible and creepy and if you want to have a bad time you should go look up some pictures online. One of the possible late stages of the disease is neurosyphilis, an infection of the central nervous system. Patients affected by it basically would usually suffer from meningitis and progressive brain damage. What made the experience even more unpleasant were mental problems, from dementia to schizophrenia, depression, mania, or delirium, all caused by the bacteria wreaking havoc. Ultimately it is fair to say that affected patients had a bad time and in the end they would die without doctors being able to help them, other than to try to alleviate their suffering. But they did observe that in some cases patients that suffered from unrelated, very high fevers would actually end up being healed. So naturally, a few doctors

While not all mechanisms and effects on the immune system are known, generally the Innate and Adaptive Immune System work better through higher temperatures from fever in a variety of ways. Neutrophils are recruited faster, Macrophages and Dendritic Cells get a bit better at devouring enemies, Killer Cells kill better, antigen-presenting cells get better at presenting, T Cells have an easier time navigating the blood and lymph system. Just overall fever seems to activate the immune system to improve the ability to fight pathogens.

How exactly does the actual temperature increase stress out pathogens and make our cells better at fighting them? Well, it all has to do with the proteins inside cells and how they work. To put it in a simplified way: Certain chemical reactions between proteins have a sort of optimal zone, a temperature range in which they are most efficient. By increasing the temperature in the body during fever, pathogens are forced to operate outside this optimal zone. Why does this not affect your cells but even helps them? Well, as we alluded to earlier, your animal cells are larger and more complex than for example bacteria cells. Your cells have more sophisticated mechanisms that protect them from higher temperatures, such as heat shock proteins. Also, your cells have more redundancies, if one of their internal mechanisms is impaired, they probably have alternative mechanisms that can take over. This is also the reason fever is helpful to your immune cells, since they can handle the heat, they can make use of the effect that higher temperatures tend to speed up certain reactions between proteins. So the complexity of your cells, in contrast to many microorganisms, makes them not suffer from fever but instead work more efficiently. Of course there is also a limit on how hot we can get and for how long before our systems break down too.*

---

began experimenting with pyrotherapy, a version of treatment by causing fever, and began injecting syphilis patients with malaria. This sounds horrible at first but it was a pretty acceptable risk—the patients would die anyway and malaria could already be treated at the time. Malaria was a prime candidate because it caused high fevers over a long and sustained time and basically cooked out the syphilis bacteria that just could not stand the heat. Indeed the treatment was so effective that it was honored with the Nobel Prize in Medicine in 1927. The onset of antibiotics made the treatment obsolete in the 1940s, making this story one of the great footnotes in medical history.

* This seems to be true for most animals. For example, lizards that were held in terrariums with higher temperatures had a higher chance of surviving an infection than lizards that were held in

Back at the escalating battlefield, Dendritic Cells swallow and scan fluids and detritus and are picking up influenza viruses. They also do get infected by them but are way more resilient than epithelial cells and keep operating, which will become important later on. Their role is super important because without the Adaptive Immune System your body has a really hard time with virus infections, especially such effective pathogens as influenza. Until it shows up though, the efforts are only delaying the infection, not stopping it, and so the virus spreads and infects more and more cells.

An Aside The Difference Between the Flu and the Common Cold

THE FLU GENERALLY FALLS INTO THE CATEGORY OF *ACUTE UPPER RESPIRA-tory tract viral infections,* which are the most common types of disease humanity has to deal with. What makes them really annoying to talk about is not only their super-handy name, but that it can mean a lot of different things on a wide spectrum. On the one end we have the common cold, a disease that even a healthy adult gets two to five times a year, and children up to seven times, and that is, everything considered, pretty harmless.*

The common cold can be so mild you don't even notice it, or it can be pretty unpleasant. The symptoms vary from none at all to a headache, sneez-

---

colder environments. And there have been a number of similar experiments with fish, mice, or rabbits and even some plant species. Transforming our bodies into a hot ecosystem just seems to be a good defense strategy against the intruders of the microworld. Interestingly, animals who can't regulate their body temperature like us mammals can, so-called "ectothermic," or "cold-blooded," animals like lizards or turtles have behavioral fever. Which means that if their immune cells release certain cytokines, they look for a hot place, like a rock that has been in the hot sun for a long time, and rest there for a while. They basically grill themselves to increase their body temperature to a point where the pathogens inside them have a really bad time.

* You know the people who use moments like this to sharply inhale before they let you know they never get sick even though you didn't ask them? Or that they haven't been sick in years because _ (enter reason that makes no sense) _ ? Rest assured: Everybody gets sick, common cold infections can just be pretty mild or we selectively remember the times when we feel fine. The best way to react to this kind of outburst is to nod politely and change the subject.

ing, chilliness, a sore throat, an obstructed nose, coughing, and a general malaise.

In the case of the flu, fever and other symptoms usually hit you like a freight train. You feel fine, maybe a tiny bit under the weather, and then *boom,* suddenly you feel actually really sick and weak as you burn up. A proper influenza A infection comes with a whole trove of nasty symptoms. Aside from a high fever, you feel extremely tired and weak, your head hurts, which makes thinking or reading hard, your throat is sore and you have to cough intensely. As if this would not be enough, as the day moves on your whole body begins to ache and hurt. The pain seemingly comes right from your muscles in your arms and legs. Most of these symptoms can be caused by other infections, they are not technically unique to the flu, which can make it hard to discern, even for doctors sometimes.

There is the common wisdom that the color of your snot can tell what kind of infection you have and if it is just a cold or a flu, but that is not true: the color just tells you how severe the inflammatory reaction inside your nose is, not what caused it. The more colorful, the more Neutrophils have given their life.

Think about this for a moment—with every sneeze, you get rid of not only thousands to millions of viruses or bacteria, but also your own cells that fought bravely and died in the process. There even might be Neutrophils still alive when you blow your nose into a tissue. A sort of sad fate, like an astronaut ejected into space. Fighting for you with all their might, only to get discarded with the enemy, ending up in a trash can. A truly horrifying fate, and if your cells were conscious this would be a pretty sad way to end their lives.

After you have spent your Saturday morning being a whiny baby and being stubborn about enjoying your weekend, the influenza A infection finally hits you for real. You begin to feel worse and worse, you get hot and weak, and all your symptoms scale up. It is no longer possible to ignore this, you are properly sick. You crawl back into bed and have no other choice than to go through this now, there is nothing you can do against the flu but to rely on your immune system working properly. Well, at least this also means you can skip work for a week or two, you think before you slide into a fevery drowse.

Within three days after the initial infection with influenza A, the replication of the virus infection peaks as the innate immune system is catching and killing as many viruses as possible. Still, the majority of the viruses are safely hidden inside infected cells, doing their dirty parasitic work in the dark behind membranes. If the fight continues this way, the virus will not be removed, there is no way around it. Since viruses spend most of their time inside infected cells, it is simply too hard to catch all of them when they float from cell to cell. If your immune system could fight viruses only when they were outside cells, they would be nearly unbeatable and humans might not be around today.

The best way to kill a lot of viruses is to destroy infected cells, and the viruses inside them. Let us pause for a moment to appreciate the magnitude of what we are talking about here. Your immune system needs to be able to kill your own cells. *Your immune system has an actual license to kill you.* As you can imagine, this is an extraordinarily dangerous power that carries an extreme responsibility; just imagine what would happen if these cells go wrong, they could decide to kill healthy tissue and organs. And indeed, this does happen to millions of people each day and is called autoimmune disease, which we will get to know much more intimately later on. So, how does your immune system do this without causing horrible damage?

30 The Window into the Soul of Cells

REMEMBER IN THE CHAPTER "SMELLING THE BUILDING BLOCKS OF LIFE," we learned that cells can smell their environment and recognize intruders and their excretions with toll-like receptors that can recognize the shapes of different enemy molecules. They do this so your soldier cells can detect enemies and kill them efficiently. While this is nice and all, this leaves still a very significant blind spot, the insides of infected or corrupted cells.

And being able to tell if a civilian cell should be destroyed is not relevant only for viral infections. Some species of bacteria, like *M. tuberculosis*, actually invade your cells and hide from your immune system while they eat their victims from the inside out. Then there are cancer cells that usually appear inconspicuous from the outside while being broken from the inside. Cells that are infected or corrupted need to be identified so they can be killed before they can cause large-scale damage, be it by spreading a pathogen or growing into a tumor. And of course, who could forget protozoa, our single-celled "animal" friends, like the trypanosoma that causes sleeping disease or plasmodium that causes malaria and kills up to half a million people each year.

So, to detect the danger of these corrupted cells, your immune system has developed an ingenious way that makes it possible for your cells to look inside other cells. In a nutshell: They do this by bringing the insides of cells to the outside. Wait, what? How does this work?

To explain exactly how, a short reminder about the nature of cells might be useful here: Cells are complicated protein machines that constantly have to rebuild and break down structures and different parts inside themselves. They are filled up with millions of different proteins, with many different jobs and functions, that work together in a beautiful concert of life.

The conductor of this concert is the DNA in the nucleus, and the arms of this conductor are the mRNA molecules that relay commands that proteins need to be made. But these proteins are more than just materials and parts. They tell a story. A story of what is going on inside a cell. If you could see a cross section of all the proteins of a cell, you could see what it is doing, what kind of stuff it is currently building, what notes the conductor wants the orchestra to play. And of course, if there is something wrong.

If for example a cell is making virus proteins, then it is obviously infected by a virus. Or if a cell is broken and turns into cancer, it will start producing faulty or abnormal proteins.*

But your immune cells can't look through the solid membrane of your cell to check what kinds of proteins are being manufactured and if everything is all right. Nature solved this differently: It brings the story the proteins tell from the inside to the outside by using a very special molecule that works like a display window.

This molecule has one of these horrible names of immunology that might feel very familiar to you:

The *Major Histocompatibility Complex class I molecule*, or in short *MHC class I molecule*. You might have guessed that this molecule is closely related to the MHC class II molecule that we got to know in depth earlier. Here immunology has chosen to be extra confusing and annoying: The two MHC molecule types are crucially important and they have *fundamental* differences. MHC class I molecules are display windows. MHC class II molecules are hot dog buns! Very different things, annoyingly similar names!

First of all, just like the MHC class **II** molecule, the job of an MHC class **I** molecule is to *present antigen*. The extremely important difference between both molecules is: Only *antigen-presenting cells* have MHC class **II** molecules.

---

* What is an abnormal protein, you ask? For example certain proteins are made only when you are an embryo inside your mother's womb. Some of these proteins make it possible for embryonic cells to grow and divide rapidly, something you need at this early stage of life but is detrimental to you as an adult. The building instructions for these proteins are still part of the DNA in adult cells—although they are not used anymore. There is a whole library of proteins like these and their presence, in anything but embryos, tells the immune system that something is off. So these proteins are not technically faulty because they are serving your tumor well, but they are definitely abnormal and thus a sign of danger for your body.

MHC class I: The Window into the Cell

The MHC class I molecule showcases random proteins from inside the cell to the outside world. This way, things like virus infections become visible to the outside.

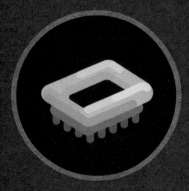

MHC class I molecule

Displaying an antigen

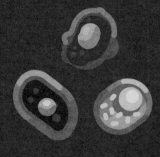

Present on every cell of your body with a nucleus.

MHC class II: The Hot Dog Bun

The MHC class II molecule presents antigens to other immune cells to activate or simulate them.

MHC class II molecule

Presenting an antigen

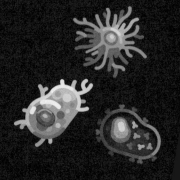

Present only on Dendritic Cells, Macrophages, and B Cells

This includes Dendritic Cells, Macrophages, and B Cells—all of them immune cells!

This is it—no other cell is allowed to have an MHC class II molecule.*

In contrast to that: *Every cell of your body that has a nucleus (so not red blood cells) has MHC class I molecules.* OK, why is that and how does this work?

As we said before, cells are constantly breaking down their proteins so their parts can be recycled and reused. The crucial thing here is that while this recycling happens, your cells pick a random selection of protein pieces and transport them to their membranes to display them on their surfaces.

The *MHC class I molecule* showcases these proteins to the outside world, just like a fancy display window would showcase a selection of the items the inside of a store has to offer. This way, the protein story of what is going on inside the cell can be told to the outside. To make sure the story is always up-to-date, your cells have many thousands of display windows—or many thousands of MHC class I molecules—and each one gets refreshed with a new protein, about once a day. Every cell in your body that has a nucleus and protein-production machines does this constantly. So your cells constantly showcase what is going on inside them, to assure the immune system that they are fine. As we will learn in the next few chapters—there are cells going through your body right now, checking these windows on random cells, making sure there is no funny business going on inside them.

Think about how genius of a principle this is and how many problems this solves. In the case of our influenza A infection the mechanism works like this: Remember that the first thing the viruses did when they successfully invaded your cells was to take over the production sites of the cell. They used the tools and resources of the cell to make virus protein parts, virus antigens.

Automatically, as a sort of background noise, some of these virus antigens were picked up and transported to the *display windows, the MHC class I molecules,* on the outside of the cell. This way the cell clearly signaled not

---

* Hey, how about an immediate exception? There is one more type of cell in your body that needs MHC class II molecules: The teacher cells in your Thymus because they need them to educate your Helper T Cells and make sure they can recognize MHC class II molecules properly!

only that it is infected but also by whom—although the enemy is hidden inside and invisible, its antigens are not!

Because all of your cells constantly display proteins in their MHC class I molecules, infected cells present their insides to the outside world, even if they don't "know" that they are infected! The display-window thing is an automated process that always happens in the background as part of the normal life of your cells. If an immune cell wants to check if a cell is infected, it simply can move closer and peer into the little "windows" to get a snapshot of the inside. If it recognizes things in the windows that should not be inside the cell, the infected cell will be killed.

Even better, the number of MHC class I molecules is not fixed. One of the most important things that happens during the chemical warfare triggered by interferon is that cells are stimulated and ordered to make more MHC class I molecules. So in case of an infection, interferon tells all cells in the vicinity to build more windows and become more transparent, to tell more of their internal protein story and be more visible to the immune system.

Another thing that is special about your cell display windows is that they are a badge of your individuality. We already mentioned in the first part that the genes that code for MHC molecules, class I and II, are the most diverse genes of the human species. If you don't have an identical twin, it is very likely that your MHC class I molecules are *unique to you* specifically. They work the same in all healthy humans, but the proteins that make up the molecules have hundreds of slightly different shapes and they are a tiny bit different from person to person.

This is incredibly important and unfortunate for one thing though: Organ transplantation. Because the MHC molecules are the place where your immune system can recognize that a cell on an organ that a generous person donated to you is not actually you. It is not *self*, it is *other*. And once *other* is recognized, your immune system will attack and kill the organ. Something that makes this scenario even more likely is the nature of organ transplantation.

An organ that is transplanted had to be taken out of another living being—to do that it had to be separated from it. Usually with sharp tools. This whole process is likely to have caused small wounds—what do wounds inside the body trigger? Inflammation, which then attracts the Innate Immune System. And if things go wrong, the Adaptive Immune System gets called right to the edges of the new life-saving organ and can call in more cells that check out the display windows only to find that they are not yours.

This is the unfortunate reason that after you receive a donated organ, you need to be on strong medication that suppresses your immune system for the rest of your life. To minimize the chance that your immune cells find the foreign MHC class I molecules and kill the cells carrying them. But of course this will leave you much, much more vulnerable to infections.

When your immune system evolved hundreds of millions of years ago it really could not have seen coming that at some point some species of apes would invent modern medicine and start transplanting hearts and lungs. But we are getting distracted. Back to the MHC class I molecule, the window into the cell. Let us meet one of your most dangerous cells that entirely depends on the display window. A brutal killer from the Adaptive Immune System that is one of your strongest weapons against viruses. The Killer T Cell, the murder specialist of your body.

31 The Murder Specialists–Killer T Cells

KILLER T CELLS ARE THE SIBLINGS OF HELPER T CELLS BUT THEIR JOB IS very different. If the Helper T Cell is the careful planner that makes smart decisions and shines through its ability to organize, the Killer T Cell is a dude with a hammer that bashes heads in while laughing maniacally. "Killer" T Cell is a perfect name considering what it does: *It kills*, efficiently, fast, and without mercy.

Around 40% of the T Cells in your body are Killer T Cells and, just like their Helper T Cell siblings, Killer T Cells come with billions of possible different and unique receptors for all sorts of possible antigens. They too have to pass the education of the Murder University of the Thymus before they are allowed to enter general circulation.

Just like the Helper T Cells need hot dog buns to recognize antigens (MHC class II molecules), Killer T Cells depend on the display windows (MHC class I molecules) to get activated.

So how would this work in our influenza A infection?

Think back to the battlefield, where millions of viruses were killing hundreds of thousands of cells. Dendritic Cells collected samples made of the detritus and viruses floating around at the battlefield, then they ripped them into antigens and presented them in the hot dog buns, their MHC class II molecules. But this would activate only Helper T Cells and not be useful for our Killer Cells. Here things become a bit complicated because there are still a lot of open questions about the exact mechanisms but the details are not that important right now.

All you need to know is that there is a thing that Dendritic Cells do called *cross-presentation*, which enables them to sample virus antigens and to present some of them in their MHC class I molecules, in their display windows,

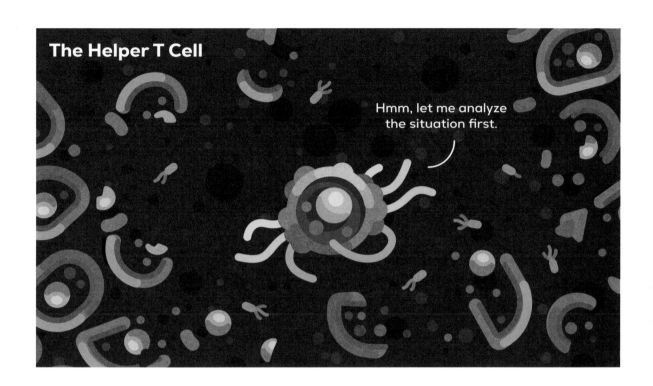

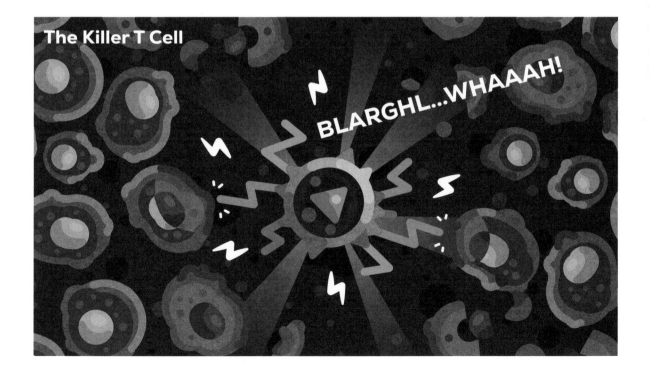

even though they have not been infected by a virus. So Dendritic Cells are able to activate Helper T and Killer T Cells at the same time, by loading up hot dog buns and display windows with antigens.*

You can imagine how Killer T Cell activation happens now. Dendritic Cells covered in antigens from dead enemies that are presented neatly in the hot dog buns and covered in virus antigens that are presented in the display windows arrive in the lymph node and move to the T Cell dating area. Here they look for a virgin Killer T Cell that is able to recognize the virus antigen in its display windows.

These Dendritic Cells that are charged up with the snapshot of a battlefield from a virus infection are basically able to call for three different types of reinforcements: They activate the specific Killer T Cells that kill infected cells, and they activate Helper T Cells that help out at the battlefield and Helper T Cells that activate B Cells to provide antibodies. And all that from one Dendritic Cell, which arrived with all the intel and antigens the adaptive immune system could ever wish for.

This is also important for a second reason, for to truly awaken, Killer T Cells need a second signal. As you might imagine, Killer T Cells are a very dangerous bunch that you wouldn't want to activate by accident. So similarly to B Cells, their full activation requires a two-factor authentication. A Killer T Cell that was activated only by a Dendritic Cell will make a few clones of itself and can fight, but it is a little sluggish and it will kill itself rather quickly.

The second activation signal comes from a Helper T Cell. So this is again the two-factor authentication we got to know with B Cells: To truly activate the strongest weapons of your adaptive immune system both the Innate and the Adaptive immune system need to "agree," they both need to give their permission.

Only if a Helper T Cell has been activated before by a Dendritic Cell and then goes on to restimulate the Killer T Cell can it live up to its full potential.

---

* Another way the Dendritic Cell can activate a Killer T Cell is by straight up being infected by the virus itself. Just like a regular cell, the Dendritic Cell presents samples of the virus in its MHC class I molecules and can still tell the adaptive immune system: "Look, there is a pathogen that infects cells, even I am infected, mobilize special forces especially for that type of enemy." To enhance the chance that this happens, Dendritic Cells that sense the chemical warfare triggered by virus infections massively produce more display windows, to become super, extra transparent.

Cross Presentation

The Dendritic Cell is able to present antigens in both MHC molecules. This way it can activate Helper T Cells and Killer T Cells simultaneously.

MHC class II

Helper T Cell

Killer T Cell

MHC class I

When truly activated, the Killer T Cell rapidly proliferates and makes lots and lots of clones of itself that eventually move to the battlefield to dish out a lot of death.

About ten days after you caught the infection in the break room you are still pretty sick. Your immune system has been fighting but it also made you feel horrible in the process and the infection is still going strong. Around this time, the Killer T Cells finally arrive in your infected lung. They squeeze by Macrophages and dead civilian cells and slowly and carefully move from cell to cell to scan them for infection. They basically press their face against the face of a civilian and take a close and deep look into the many display windows on their surface, carefully scanning the story of the insides. If they don't find antigens they can connect their T Cell receptors to, nothing happens and they move on.

But when a Killer T Cell finds a cell that has virus antigen in its display window (MHC class I receptors) it immediately issues a special command to the cell: "Kill yourself but be very clean about it." Not with aggression or anger, but matter-of-factly and with dignity, if you want to anthropomorphize this process. The death of the infected cell is a necessity, a fact of life, and it is important that it occurs properly. This is one of the key parts in the response to virus infections:

It is very important how an infected cell kills itself. If the T Cell, for example, used just chemical weapons and threw them around, like the way Neutrophils do it, they would rip their victim open and make them burst. Not only would this release the cell's guts and insides and cause harsh inflammation reactions, it would also release all the viruses inside the infected cell that have been made until this point.

So instead, the Killer T Cell punctures the infected cell and inserts a special death signal, that is conveying a very specific order: *Apoptosis*, the controlled cell death we mentioned before already. This way the virus particles are neatly trapped in tiny packets of cell carcass and unable to do further damage until a hungry Macrophage passes by and consumes the remnants of the dead cell. This process is extremely efficient and the virus count drops harshly as thousands of Killer T Cells move through the battlefield, checking every cell they meet for infection, in a process that is called "serial killing." Yes, this really is what it is called, praise where praise is due, immunologists

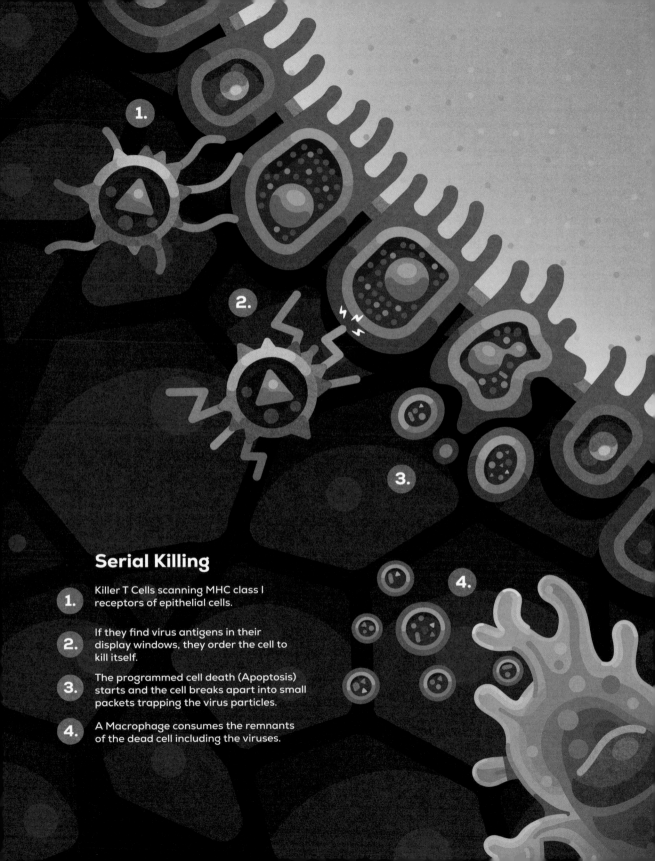

Serial Killing

1. Killer T Cells scanning MHC class I receptors of epithelial cells.

2. If they find virus antigens in their display windows, they order the cell to kill itself.

3. The programmed cell death (Apoptosis) starts and the cell breaks apart into small packets trapping the virus particles.

4. A Macrophage consumes the remnants of the dead cell including the viruses.

nailed this term. Millions of viruses are destroyed before they get the chance to infect more victims. But also hundreds of thousands of infected civilian cells are ordered to kill themselves this way. No, the immune system has no chill at all, it does what it needs to do.

Unfortunately there is a huge flaw in this system: Pathogens are not stupid and they found ways to destroy the display windows and therefore hide themselves from the immune system, from Killer T Cells. Many viruses force infected cells to stop making MHC class I molecules, effectively destroying this strategy.

So in this case, are you doomed?

Well, of course not because your ingenious defense network has an answer, even for this case.

And it has one of the best names in all of immunology: Meet the **Natural Killer Cell.**

32 Natural Killers

NATURAL KILLER CELLS ARE CREEPY FELLOWS.

They are related to T Cells, but when they grow up they leave the family business and join the Innate Immune System. Think of them as children from a family that has served as fighter pilots for generations, who defy tradition and instead become army grunts. Refusing to follow the footsteps of their family and the more prestigious role in the defense and instead purposefully seeking fulfillment in the more hands-on, brutish ground combat.

Natural Killer Cells are sort of inconspicuous fellows that nonetheless are one of the few cells with an official license to kill your own body cells. In a way you can imagine them as inquisitors of the vast empire of your immune system. Always looking for corruption and able to act as a judge, jury, and executioner. In a nutshell, Natural Killer Cells hunt two types of enemies: cells infected by viruses and cancer cells.

The tactic Natural Killer Cells employ is nothing short of genius.

Natural Killer Cells do not look inside cells. Even if they wanted to, they couldn't—they have no way to look into the display windows, the MHC class I molecules, and read the story of the inside of the cell.

No, instead they do something different: *They check if a cell has MHC class I molecules. Nothing more, nothing less.* This is solely to protect against one of the best anti–immune system tactics virus and cancer cells have. Generally cells that are either infected or unhealthy do not show MHC class I receptors in order to hide what is going on inside them. Many viruses force infected cells to stop showing them as part of their invasion strategy, and many cancer cells just stop putting up display windows, thus making them invisible to the antiviral immune response that we have shown thus far.

The Adaptive Immune System is now suddenly harmless to these cells. In a very real way, without their display windows, the infected cells go dark and become impossible to detect. This is quite an effective tactic if you think

211

about it—all a virus or cancer cell needs to do is to stop making a single molecule and boom, the extremely powerful response of the body is helpless.

So the Natural Killer Cell just checks for one thing: Does a cell display a window? It does? "Great, please carry on, sir!" It doesn't? "Please kill yourself immediately!" That's right. The Natural Killer Cell is specifically looking for cells that do not share information about their insides, that don't tell stories. The Natural Killer Cell removes the fatal flaw that otherwise could so easily become deadly. This is such a simple principle but has such a powerful effect.

While the rest of the immune system looks for the *presence of the unexpected*, the presence of something *other*, Natural Killer Cells look for the *absence of the expected*, the absence of *self*. This principle is called *"The Missing-Self Hypothesis."*

The mechanism of how it works is equally fascinating as the strategy itself: The Natural Killer Cell is always "on"—when it approaches a cell it is doing so with the "intention" of killing. To prevent them from killing healthy cells, they have special receptors that calm them down, an inhibitor. A large stop sign receptor. The display window, the MHC class I molecule, is this stop sign, and it fits right into this receptor.

When Natural Killer Cells check out a civilian cell for infection or cancer, if they have plenty of MHC class I molecules, as most healthy cells do, the inhibitor receptor is stimulated and tells the Natural Killer Cell to chill. If the cell does not have enough MHC molecules though, there are no calming signals and the Natural Killer Cell, well, kills.

Killing in this case means that it orders the infected cell to kill itself through apoptosis, the regular and orderly cell death, which traps viruses inside the carcass. So Natural Killer Cells are a bit like nervous agents who are walking through a city, randomly approaching civilians. Instead of saying hi, they put a gun against your head and wait a few seconds. If you don't show them your passport quickly enough, they pull a plastic bag over your head and shoot you in the face.

Natural Killer Cells are really quite scary guys.

OK—but does this mean that Natural Killer Cells are useless if an enemy does not try to hide its MHC class I molecules? Not at all, there is a bit more

Healthy cell

MHC class I

Infected cell

Natural Killer Cell

Natural Killer Cells

Natural Killer Cells arrive at the
battlefield about 2-3 days after the
infection started. They check if a cell is
stressed or is showing no MHC class I
molecules. If they don't, they order the
cell to kill itself.

to this story but the most important part was the display window. Natural Killer Cells are looking for stress, looking for cells that are unwell. Not just during an infection, by the way, right now in this second, millions of these cells patrol your body and check your civilian cells for signs of stress and corruption, cells that are on the verge or have already turned into cancer.

Cells have a number of ways to communicate to their environment how they are and if things are fine and they are well. And there are subtle ways they can express their inner state that are not quite as obvious as asking for help. Ways not as obvious as display windows.

Imagine that one of your friends had a bad time in their lives but they were not ready to tell anybody, but you still might notice that they smile less, or tend to have a worried expression or don't react as enthusiastically to good news as you would expect them to. Because you know them well, you would pick up these signals and could ask them in a quiet moment if everything is OK and if you could help.

In a sense this is something Natural Killer Cells can do too with your civilian cells. If a cell is under a lot of stress—which in this context means that something is affecting the complex cellular machine, made of millions of proteins, in a negative way, say for example because a virus is interrupting the machinery or a cell is becoming cancerous and not working as it should— the cell will express certain stress signals on its membrane.

The details of these stress signals are not important, imagine them like the face of your friend getting more and more unhappy looking. More stress, more unhappy wrinkles. Natural Killer Cells can detect these stress signals and they too can take the cell aside to have a nice talk with them. The difference for you and your Natural Killer Cells is that they do not want to talk it out and ask if there is anything they can do to help. If Natural Killer Cells detect too many stress signals, they shoot the poor stressed-out cell in the head. So if there were human-size Natural Killer Cells, it would be important to smile a lot around them!

And this is not all! Remember IgG Antibodies, the all-purpose Antibody that comes in different flavors? Natural Killer Cells can interact with them too!

Specifically in case of an influenza A infection they work together splendidly! Remember that the viruses budded from the infected cell, taking part of its membrane with them? Well, this process is not instantaneous but takes a while—enough time for IgG Antibodies to grab onto a virus before completely detaching. Natural Killer Cells can connect to these Antibodies before the virus particles detach and order the infected Cell to kill itself.

Infected Cells are just not safe from Natural Killer Cells.*

OK, now that we have met every major player of your antiviral defense, let us bring them all together!

---

* Except, of course, if they are red blood cells. As we said before, they are the only cells in your body that don't have MHC class I receptors, no display windows. This is what happens in malaria—the parasite plasmodium infects red blood cells and Natural Killer Cells cannot check these cells for windows and have to do something else to combat this infection.

33 How a Viral Infection Is Eradicated

WHEN WE LEFT THE BATTLEFIELD THE LAST TIME THINGS WERE TURNING gruesome. Millions of cells were dying and your innate immune system was desperately and pretty unsuccessfully trying to rein in the rapidly spreading infection. Numerous chemical signals were flooding your body, requesting a change in temperature and making your body burn up with a high fever, kicking your immune system into a higher gear and making it fight harder.*

All sorts of systems began to wake up, producing more mucus and making you cough violently, removing millions of virus particles from your body but also making you highly infectious. The onslaught of battle chemicals, cytokines, and dead or dying cells makes you exhausted and causes all sorts of unpleasant body sensations.

But all that was just to buy time.

It takes about two to three days until Natural Killer Cells show up and begin to alleviate your desperately fighting immune soldiers. They flood the tissue and begin killing infected epithelial cells, especially the ones that were manipulated by the influenza A virus to hide their display windows, their MHC class I molecules, but not exclusively them. The more stressed and desperate infected cells get mercifully finished off, to end their suffering but also to prevent them from causing further harm.

The arrival of the Natural Killer Cells is a noticeable relief for the defense of the battlefield as they are actually causing a real dent in the number of

---

* If a fever reaches 104°F (40°C) it becomes dangerous for humans and you should immediately seek out medical attention. Around 107.6°F (42°C) your brain is starting to get damaged but this is very rare and seldom a side effect of disease, as the body will usually stop itself from getting too hot.

infected cells. But even these merciless and effective killers are not enough to end the infection. Even they are basically just buying time, although much more successfully than the Macrophages, Monocytes, and Neutrophils.

While this was going on thousands of Dendritic Cells had sampled the battlefield and picked up viruses, ripping them into pieces that are put into their MHC class I molecules (and their MHC class II molecules). They made their way to the lymph nodes and activated Killer T Cells and Helper T Cells that then went on to activate B Cells and order Antibodies.

And now, about a week after you collapsed into bed, your heavy artillery finally arrives.

Killer T Cells flood into your lungs in the thousands, armed with receptors that recognize influenza A virus antigen, moving from cell to cell and taking them into a warm embrace, taking a deep look into the MHC class I windows, listening to the protein story that the cells tell. If they detect virus antigens they order the infected cells to kill themselves. Macrophages work overtime to eat up all their dead friends and foes.

Millions and millions of Antibodies move in to eliminate the viruses outside of cells and stop them from infecting more cells. Through the magic of the dance of the B and T cells, a number of different types of Antibodies were made that attack the virus on different fronts.

Neutralizing Antibodies neutralize the virus by connecting firmly to the structures that the virus used to gain access into epithelial cells. Covered by dozens of Antibodies that block the virus from entering cells, it is now nothing more than a useless and harmless bundle of genetic code and proteins that will eventually be cleaned up by your Macrophages.

Other Antibodies can be very specific and block the virus in a variety of interesting ways. For example, there is a virus protein called viral neuraminidase that enables new viruses to be released from an infected cell. As we explained earlier, influenza A viruses bud from infected cells, taking a bunch of the membrane of their victims with them. Antibodies can connect to viral neuraminidase during this budding process and effectively disable it. So you have an infected cell with lots of new viruses on its surface that are unable to detach and infect new cells, basically being stuck in place like flies on one of these sadistic glue traps.

Antibodies and Killer T Cells in concert really do the trick and the num-

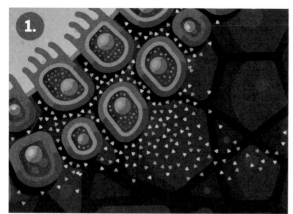

1. Your respiratory mucosa got infected by a virus, which multiplied millions of times. Your infected epithelial cells send out Interferons to sound alarm.

2. After 2-3 days Natural Killer Cells arrive and start killing infected and stressed cells.

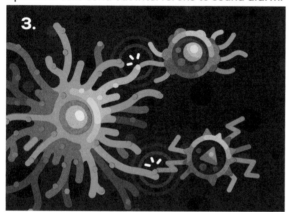

3. Dendritic Cells sample the battlefield and move on to the lymph nodes to activate both Killer and Helper T Cells.

4. Activated Killer T Cells move in and order infected cells to kill themselves, Macrophages clean up the remnants.

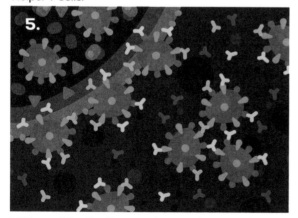

5. Millions of Antibodies sent from activated B Cells clump up viruses, blocking them from entering other cells or trap them on the host cell membrane.

6. The battle is won and most viruses cleaned up. Now it's time to shut down your immune system again, before it will do terrible damage.

ber of viruses in your lungs collapses rapidly. Over the next few days the combined symphony of your immune system eradicates the majority of the infection and begins a great cleanup on the battlefield. It seems like the war is already over, but this is not quite true.

In contrast to our first story in the bacteria section of this book, we are dealing with a different kind of immune response here. It is more global, touching way more systems, organs, and tissues, and the infection is way more dangerous. And while you lie in your bed and feel horrible, it is important to remember that the symptoms you feel are mostly created by your own immune system to clear out the infection. If these countermeasures are used too freely, your immune system can do immense and terrible damage to you, much worse than even the influenza A virus could do.

So there is an intense need to downregulate the immune response again, to have it strike with just the exact amount of vigor, and to shut it down as soon as it is not needed. To get back to *homeostasis*.

An Aside Why Don't We Have Better Medication Against Viruses?

ONE THING YOU MIGHT HAVE ASKED YOURSELF, ESPECIALLY IN THE CONtext of the global COVID-19 pandemic: why don't we just have good medication against viruses? Why is it that we have so many different antibiotics that protect us from most types of bacteria, from the plague to urinary tract infections to blood poisoning, but nothing really great against the flu, the common cold, or the coronavirus? Well, here we encounter a fundamental problem: *viruses are too similar to our own cells*. Wait. What? Well, they are not similar in the sense that a virus is similar to a cell, but in the sense that viruses mimic or work with your own parts.

In our modern times, we are used to the idea that medicine will figure things out. We are largely free from dangerous infectious diseases in developed countries and so the fact that we don't have effective medication that helps with virus infections is a bit irritating to learn. Why is that? It is best to demonstrate this with bacteria, living beings that we diverged from a long, long time ago.

Let us use this moment to explain how antibiotics work. Like Prometheus who stole fire from the gods and gave it to humanity and made it more powerful, scientists stole antibiotics from nature to make it live longer. In the wild, antibiotics are typically natural compounds that microbes use to kill other microbes. Basically the swords and guns of the microworld. The first successful antibiotic, penicillin, is a weapon from the mold *Penicillium rubens* that works by blocking bacteria's ability to make cell walls. As a bacterium tries to grow and divide it needs to produce more cell walls and *Penicillium* has a shape that interrupts this construction process and prevents the bacteria from making more of itself. The reason why you can safely be treated with penicillin is that your cells don't have cell walls! Your cells are lined with membranes, which is a fundamentally different structure, so the drug doesn't do anything to your cells.

Another antibiotic you may have heard of is Tetracycline, which was stolen from a bacteria called *Streptomyces aureofaciens* and works by inhibiting protein synthesis. If you cast your mind back to how proteins are made, you'll recall something called a ribosome. Ribosomes are the structures that turn mRNA into proteins. So they are fundamental to the survival in both human and bacterial cells, because without new proteins, a cell must die. Human and bacterial ribosomes are different in shape and as a consequence of this difference, although they basically do the same thing, Tetracycline is able to inhibit bacterial ribosomes and not yours.*

So in a nutshell, bacterial cells are very different from your cells. They use different proteins to stay alive, they build different structures like special walls, and they reproduce differently than your cells. And some of these differences present great opportunities for us to attack and kill them. A good drug is basically a molecule that connects to the specific shape of a part of an enemy (not too dissimilar to an antigen and a receptor!) that is not present

---

* Hey, time for an exception! You do have bacteria-like ribosomes in almost all of your cells. Remember that your mitochondria, the powerhouse of the cell, were ancient bacteria in the past. Because they have kept their own ribosome Tetracycline can also disrupt them, which is not great and leads to pretty unpleasant side effects. One more reason why we need a diverse set of antibiotics.

in your body. In principle, this is how many drugs and antibiotics work. They attack a difference in shape between bacterial and human parts.

OK, yeah, sure, so what is the big deal here, why don't we have drugs against viruses? Well, we have them. Actually we have thousands of different drugs that can treat viral infections. The only catch is that *most of them are pretty dangerous and sometimes even deadly to us.* Many are really more like a desperate last resort, the kind of thing that only gets used when the life of the patient is already on the line.

Think about the nature of a virus. Viruses can be attacked in two places, outside of your cells and inside of cells. If you want to attack them outside of your cells, then you basically have to attack the proteins they use to connect to the receptors of your cell. The huge, monumental problem with that, is that if you do this, you may just have created a drug that will also connect to a lot of parts inside your body. *Because to connect to one of your receptors a virus needs to mimic a part of your body.* A part that fulfills some sort of vital function. If you develop a drug that attacks a virus that connects to this receptor it is likely that it will target all the parts of your body that are supposed to connect to the receptor. It is the same inside your cells—we can't make antiviral drugs that target different metabolic processes of a virus, like for example the ribosome. Because this is *our* ribosome the virus is using. A virus is in a perverse way very similar to us because it uses our own parts to make more of itself.

34 Shutting the Immune System Down

ABOUT A WEEK AFTER THE FLU HIT YOU LIKE A FREIGHT TRAIN YOU WAKE UP one morning and feel considerably better. Not over it, but better. Your temperature is lower, you have some appetite, and just overall you feel more like yourself. The next few days your job will be to rest and let your immune system clean up and wind down as you enjoy the last few days of being sick that mostly consist of watching TV and being cared for by increasingly annoyed loved ones.

This "winding down" phase is as important as activating your immune system. An active immune system causes collateral damage and uses up a huge amount of energy so your body wants it to be done as soon as possible. But then again, how crazy dangerous would it be if your immune system stopped working before a disease was beaten and pathogens could flare up again, overwhelming the retreating forces?

It needs to shut itself down at precisely the correct time, which is easier said than done if you have millions and billions of active cells fighting, without a form of central authority or conscious thought. So as with its activation, your immune system relies on self-enforcing systems to end a defense.

Activation usually begins with an initial exposure of immune cells to intruders, like bacteria, or danger signals, like the insides of dead cells. For example, Macrophages get activated when they notice an enemy and release cytokines that call up Neutrophils and cause inflammation. The Neutrophils themselves release more Cytokines, causing more inflammation and reactivating Macrophages, who continue fighting. Complement proteins stream into the site of infection from the blood, attack pathogens, opsonize them, and help the soldier cells to swallow the enemies.

Dendritic Cells sample enemies and make their way to the Lymph Nodes to activate Helper T Cells or Killer T Cells, or both. Helper T Cells stimulate the innate immune soldiers to continue fighting and create more inflammation. Killer T Cells start killing infected civilian cells with support from Natural Killer Cells. Meanwhile, activated B Cells have turned into Plasma Cells and release millions of Antibodies that stream onto the battlefield and disable the pathogens, maiming them and making them much easier to clear up. This is the immune response in a nutshell.

As more and more enemies are killed and their numbers are dwindling, fewer and fewer battle cytokines are released, because fewer immune cells are stimulated by ongoing fights.

This means no more new soldiers are called in while the old ones die off or stop fighting. The cytokines that cause inflammation are used up relatively quickly, so without new and engaged soldiers that constantly release new cytokines, the inflammation reactions begin receding naturally, which also slowly has the complement system fizzle out.

Fewer signals from the battlefield means that the activation of new T Cells first slows down and then stops, while active T Cells become harder to stimulate the longer they are active, until eventually most of them kill themselves.

No part of your immune system is made to work forever without stimulation and so if the chain of activations stops, step-by-step, the immune response winds down.

In the end your Macrophages devour and clean up the carcasses of the brave immune cells that fought so hard to eradicate the infection to protect you. And so, just as your immune system is winning, it is already shutting itself off without any sort of central planning.

Of course there are exceptions because there is one type of cell that actively switches your defenses off and calms down the immune response: *Regulatory T Cells*. They make up only about 5% of T Cells and in a sense are "opposite Helper T Cells."

For example they can order Dendritic Cells to become worse at activating the Adaptive Immune System or they can make Helper T Cells slow and tired, so they don't proliferate as much. They can turn Killer T Cells into much less vicious fighters, shut down inflammation, and make it recede faster. In a nutshell, they can end an immune reaction or just prevent it from being triggered in the first place.

Especially in your gut, Regulatory T Cells are crucial—which makes a lot of sense when you think about it: What really is the gut if not an endless, tubelike metropolitan area for commensal bacteria that your body wants to be there? It would be hugely detrimental for your health if your immune system in the gut was truly set free here. Constant inflammation and constant fighting would be the consequence. So Regulatory T Cells keep the peace. Maybe their most important job though is to act as a countermeasure to autoimmune diseases, where they prevent your cells from attacking your body.

Regulatory T Cells are one of the parts of the immune system where things become very blurry. In this book we are trying to be clear and to paint the picture of a structured and orderly system. Unfortunately there are areas where this is harder to do than in others and Regulatory T Cells is one of them. So we will not dive into any more detail here because there is a lot of complexity buried here and a lot that is not completely understood yet.

OK, so we learned how an immune response is triggered, how it clears an infection, and how it powers down afterwards, but we are still missing the last really important puzzle piece: Your long-term protection, also known as immunity. Why do you get many diseases just once in your life and what does it mean if you become "immune" to anything?

35 Immune—How Your Immune System Remembers an Enemy Forever

THINK OF THE INFLUENZA A INFECTION THAT KILLED MILLIONS OF YOUR cells in one of your most important organs and forced you to bed for two weeks. Beating such an invasion comes at a massive cost to your whole body even in our modern world, and indeed up to half a million people are killed each year by the flu. One can only imagine how dangerous an infection like this would have been to our ancestors living without the protective veil of civilization where things like safe shelter and food are no concern. Your body *really* does not want to do this again, being sick leaves you vulnerable, and in the worst case it kills you.

Remembering the enemies it fought in the past and keeping that memory alive is one of the most important abilities of your immune system. Only through memory do you become *immune*, which roughly translated from Latin means "*exempt*." So if you are immune, you are exempt from a disease. You can't be struck down by the same illness twice (of course there are exceptions, there are always exceptions . . .).

The fact that our bodies are becoming immune to diseases after contracting and surviving them is not a new idea though. Twenty-five hundred years ago when Thucydides, the first modern historian in human history, wrote down his account of the Peloponnesian War between Athens and Sparta, he observed during an outbreak of the plague, that people who survived the disease seemed to be immune to it afterwards.

Without immunological memory, you would never become immune to anything, which would be like a horrible nightmare if you think about it. Every time you beat a serious infection your body is weakened. It costs a lot of energy to make all these immune cells and repair the damage they cause, and the devastation from the pathogen itself needs to be cleared up too. Say you survive Ebola, smallpox, or the Black Death or COVID-19 or hell, just the flu, only to get it *again*, a few weeks later. How often in a row could you survive that, even if you are a healthy adult? Without immunity modern civilization with its cities and large agglomerations of people would be impossible. The danger of getting constantly reinfected with the worst pathogens in existence would just be too high.

So you have an immune memory, and it is a living thing! Or many living things, as we introduced briefly before: **Memory Cells.** Approximately 100 billion living beings, 100 billion parts of YOU, sitting all over your body, doing nothing but remembering what you went through. Isn't this a tiny bit poetic, the fact that being immune means that there is a part of you remembering your struggles and making you stronger by its presence?

Memory cells are one of the main reasons why young children often die of diseases that their parents shake off easily: There are just not enough living memories in their tiny bodies yet, and so even smaller infections can spread and become a mortal danger. Their parents, with an Adaptive Immune System that is remembering thousands of invasions, can just rely on their living memory. And likewise, as we reach old age, more and more Memory Cells stop working as well as when they were younger or just call it quits, leaving us exposed in the last phase of our lives.

To briefly refresh your memory: B Cells need two activation signals to truly activate. The first one is delivered by antigen floating through the lymph nodes, and it leads to moderately activated B Cells. But if an activated Helper T Cell joins the party, it can deliver signal number two and confirm that the infection is serious, which activates the B Cell in earnest. Now the B Cell turns into a Plasma Cell that rapidly makes a lot more of itself and begins producing antibodies. So far so good, let's add another layer of detail!

After B Cells are activated through T Cells, some of them will turn into different kinds of Memory Cells! Living memory that will protect you for months, years, maybe even your whole life.

The first group is called *Long-Lived Plasma Cells* who wander into your bone marrow and as their very creative name suggests, they live pretty long. Instead of vomiting out as many Antibodies as they can, they make them-

1 Drop of Blood

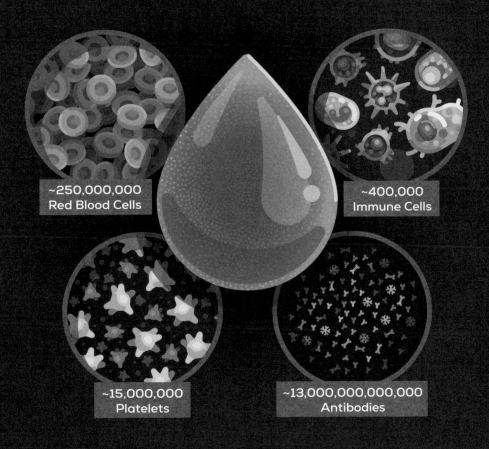

~250,000,000
Red Blood Cells

~400,000
Immune Cells

~15,000,000
Platelets

~13,000,000,000,000
Antibodies

Blood Composition:

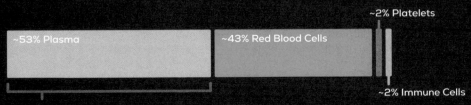

~2% Platelets

~53% Plasma

~43% Red Blood Cells

~2% Immune Cells

~92% Water
~7% Proteins (Antibodies, Complement, Albumin, etc.)
~1% Nutrients, gases, waste, etc.

selves comfortable and find a home where they will stay for months and years. From there they constantly produce a moderate amount of Antibodies. So their entire job is to make sure that specific Antibodies against enemies we fought off in the past are always present in your bodily fluids.

If the enemy shows up ever again, it will immediately get attacked by these antibodies and probably has no chance to become a real danger again. This is an extremely efficient tactic and in fact a single drop of your blood contains about 13,000,000,000,000 Antibodies. Thirteen million, million. A protein memory of all the challenges you overcame in your life.

But this is not all, there are also **Memory B Cells**. They do: Nothing. Nothing at all. Memory B Cells also settle down in your lymph nodes after they are activated, and just chill.

For years and years they are inactive and just silently scan the lymph for the antigen they remember. If they ever catch an antigen they awake suddenly and react without any sense of humor. They very rapidly proliferate and make thousands of clones of themselves that don't need Helper T Cells to properly activate but start as Plasma Cells that immediately begin to produce millions of Antibodies en masse.

This is the reason why you are immune forever against so many diseases and pathogens you encountered in your life—your Memory B Cells basically can activate directly, without going through all the complicated dances and confirmations that we showed throughout the book so far. They are shortcuts that can activate your Adaptive Immune System in a heartbeat.

What makes Memory B Cells so powerful from the start is that the fine-tuning that happens to their receptors, as we described in the chapter "The Dance of the T and the B," has already happened to them. They already went through the process and have become extremely good at making the perfect Antibodies for the pathogen. So if the intruder now attacks again, it will be confronted with Antibodies that are the most deadly to them.

In a similar vein, activated T Cells also produce Memory Cells, although with a few key differences. For one, after an infection is over, about 90% of all T Cells who fought at the site of the infection will kill themselves. The remaining 10% will turn into *Tissue-Resident Memory T Cells,* and become silent guardians. These Memory T Cells are dormant sleeper agents, that lie and wait without doing anything. If they ever spot the intruder again they will awaken and attack and activate the surrounding immune cells immediately though.

But this is not enough because this would only protect the infected area and not the rest of the body, so there are *Effector Memory T Cells*. For years they patrol the lymphatic system and your blood, not causing trouble, just looking for the antigen that once activated the cell that was their ancestor. And lastly, there are *Central Memory T Cells,* which remain stationary in your lymph nodes, doing nothing but keeping the memory of the attack. When activated they quickly produce massive amounts of new Effector T Cells that go on the attack right away.

Now, all of this is simple enough (relatively speaking). But it can't be overstated how incredibly effective memory cells are. They are so powerful and deadly that you usually will not even notice if you are reinfected by the same pathogen—even if it is a serious and dangerous one. Once your body has memory cells against an invader you are basically immune for decades, if not your whole life.

What makes your immune living memory so deadly? Well for one, they exist in much larger numbers. As we discussed before, your body makes only a few B and T Cells for every possible invader. Think back to our example of the dinner party with millions of possible guests. Our immune system cooks tried to be ready and prepared countless dishes, with every combination of ingredients. Each dish represents a unique T or B Cell with a unique receptor for one specific antigen. So when an infection first occurs, you might have only a dozen cells that are able to recognize the antigens of the enemy that invades your body.

And this makes sense because most of the many billions of B and T Cells your body makes during your lifetime will never see action. Your immune system is just trying to be prepared for every eventuality, no matter how unlikely it is. But once a pathogen with a specific antigen shows up, your Immune System knows that the antigen exists. So at this point the investment of having many more of the specific cells in store that can fight the pathogen is justified.

In our dinner party example, this is like getting confirmation about what ingredients and dishes your guests really liked! So for the future, the immune system cook can keep certain dishes in the freezer so he can serve the guests quickly if they show up again.

Just by pure numbers, the chances that if the same pathogen invades you again, one of your memory cells will be activated very early and catch the enemy very quickly are pretty high. All of these properties together make you immune against the vast majority of dangers you had to face in your past and increase the chance of your survival significantly. There are diseases though that are able to destroy your immunological memory. To kill the memory cells that defend you. It is tragic that one of these diseases is currently making a huge comeback: Measles.

An Aside What Doesn't Kill You Doesn't Make You Stronger: Measles and Memory Cells

MEASLES IS ONE OF THESE CONTROVERSIAL DISEASES WHOSE FATE IS TIGHTLY bound to the movement against vaccinations. While measles was on its way to becoming the second human pathogen to be completely eradicated since smallpox, it has made somewhat of a comeback in the last few years as more and more people decided to not vaccinate their kids against the virus.

Ironically these movements are primarily based in the developed world where people have forgotten how much of a serious disease measles still is. Globally, measles killed more than 200,000 people in 2019, most of them children, a 50% increase since 2016. Despite this sad and unnecessary rise in deaths, if you get measles in a developed country with access to good healthcare, chances are still very good that you will recover.

But there is a vicious part of measles that is not discussed as much as the disease itself: Kids who overcome a measles infection have a higher chance of getting other diseases afterwards because the measles virus kills Memory Cells. If you think that sounds a bit scary, that is the correct reaction—the virus basically deletes your acquired immunity. Let us explore how this works now that we know all the different elements of your immune system.

The measles virus is extraordinarily infectious—considerably more infectious than the novel coronavirus for example. Similar to many other viruses, measles spreads through coughs and sneezes and floats through the air in tiny droplets that stay airborne for up to two hours. If you have measles, you're so contagious that 90% of all susceptible people that come close to you will be infected just by being in your vicinity. So if you have it and other, non-vaccinated people share a subway train or a classroom with you, it is highly likely that you will infect others.

The favorite victims of measles are your T and B Cells but especially your Long-Lived Plasma and Memory B and T Cells that are vulnerable to the virus. Measles is targeting the actual, living, breathing memory part of your immune system and at the peak, millions, if not billions of your immune cells may be infected.

Luckily your immune system usually regains control of the situation and eradicates the measles virus. But the memory cells that were infected by the virus are dead and can't be brought back. Before the infection your body was full of specific Antibodies, and many of them are no longer being produced. On top of that a lot of the wandering Effector Memory Cells have died off. It is as if your immune system suffers from sudden and harsh amnesia.

So in the end, being infected with measles erases the capacity of the immune system to protect you from the diseases that you overcame in the past. Even worse, a measles infection can wipe away the protection that you might have gained from other vaccines, since most vaccines create memory cells. Therefore, in the case of measles, what does not kill you makes you weaker, not stronger. Measles causes irreversible, long-term harm and it maims and kills children.

If we lose our grip in the war on measles we will see the number of people, and especially children, dying of preventable diseases increase each year. In any case—this might be a good moment to talk a bit about a huge idea in our history. The idea to cause immunity without suffering a disease.

36 Vaccines and Artificial Immunization

As we said before, even thousands of years ago, people noticed that getting some diseases once made a patient immune to them. But it would take quite some time for the observations to turn into something actionable as people began to ask themselves if it would somehow be possible to give a mild form of a disease to a healthy person on purpose to protect them from a more dangerous infection.

Hundreds of years before humanity knew about the microworld, before anybody knew about bacteria or viruses, someone came up with the method of *variolation*: The attempt to somewhat artificially induce immunity to one of the most horrible diseases that plagued our species for thousands of years: smallpox.

In today's modern world where we are mostly being spared from outbreaks of horribly deadly diseases, it is hard to imagine how much of a scourge smallpox used to be until basically a minute ago, in human history terms. Up to 30% of all the people who contracted smallpox died, and many survivors had extensive scarring on their skin, leaving them disfigured, while as a horrible bonus, some permanently lost their eyesight. It was a family-destroying, life-ruining plague that our ancestors had little safeguards against. In the twentieth century alone, smallpox killed more than 300 million people, so the motivation to do something about this scourge was pretty high.

When, exactly, the first experiments with variolation started is not known, but it was at the very least hundreds of years ago, in medieval China. The basic idea was pretty straightforward: Take a few scabs from an infected person that only had a mild case of smallpox, let them dry out, and grind them to a fine powder. Then blow the powder up the nostril of a person you want to give immunity to. If things went well, a mild outbreak of smallpox was the consequence in the patient, who then subsequently gained immunity against harsh forms of the disease for the future.

Although a tiny bit disgusting, in times where people had no real recourse against diseases, this method was the best protection against smallpox available and so it spread around the globe. Different areas of the world practiced variolation in different ways, using needles or small cuts to rub in scabs or pus from infected people.

Still, variolation was not without risk with as much as 1–2% of all patients going through the procedure contracting a more serious version of smallpox, with all the potentially negative consequences. But the disease was so horrible and so widespread, for so long, that many took on the risk for themselves and their loved ones. So the general idea of immunization was already around for quite some time when the first proper vaccine was developed.

The proper start of the history of vaccination was the realization that it was not necessary to variolate with the actual real smallpox disease, but much safer to use material from cowpox, which was a variant of smallpox that affected, surprise, cows. This was a truly revolutionary step and only a few years later, the first vaccine was developed that would eventually lead to the total eradication of smallpox.*

As a consequence of the success of this first vaccine, more and more vaccines were developed against different horrible diseases, such as tetanus, measles, polio, and many more.

Today, vaccines provide immunity against a whole plethora of dangerous infections by creating Memory Cells that are ready to meet a specific pathogen in case it ever shows up for real. Unfortunately creating memory cells is actually far from a trivial affair. As we discussed before, your immune system is very cautious and requires very specific signals to boot up and activate

---

* While this may sound easy and straightforward, it wasn't. It still took a worldwide vaccination program and over 200 years to finally bring smallpox to its knees. To this day, smallpox remains the first and sadly only human pathogen that humanity has wiped out completely. Smallpox no longer occurs in the wild and only remains (hopefully) safely stored in two labs, one in the United States and one in Russia.

properly. To provoke the creation of Memory Cells that stick around for years, your immune system has to go through multiple steps of escalation, with the two-factor authentication and all that jazz!

So to make a good vaccine we somehow need to safely provoke an immune response to make the immune system think a real invasion is happening so it makes memory cells—but without accidentally causing the disease we want to protect ourselves against. This is much harder than it sounds and there are a number of different ways to induce immunity in a patient, some more permanent than others. Let us take a brief look at a few different methods:

Passive Immunization—Free Fish

Imagine you found yourself in Australia, a country where people are very nice and talk funny but basically everything else is a deadly venomous animal trying to kill you.*

Now imagine you continue to show bad judgment and take a guided tour through the bush to experience nature and all that. You admire the wonderful landscape and your mind wanders off, paying less and less attention to your surroundings, when it suddenly happens—spooked by the equally unalert members of the tour that walked before you, a very unhappy and stressed-out snake decides to defend itself before one of the loud apes steps on it and quickly bites you in your ankle.

The pain is sharp and immediately your ankle swells and hurts horribly, which you communicate to the rest of the world through an appropriate amount of screaming and cursing. You are lucky that the next hospital is not that far away, you are being told as you lie in pain in the backseat of a jeep,

---

* The chances of your dying through a snakebite in Australia are actually pretty low. Only around 3,000 people are bitten by snakes each year, and on average two of these die. Still, the amount of venomous critters on this continent is way too high and no amount of real-world statistics will convince me otherwise.

and while you might not feel lucky in such a situation, you actually are because you are about to enjoy the wonders of *passive immunization*.

Passive immunity is basically the process of borrowing immunity against a disease or a pathogen from someone who survived something. Since we can't easily borrow immune cells, as your immune system would immediately recognize them as *other* and attack and kill them, we are talking about antibodies here. How does this work in the case of a snake bite that comes with horrible venom?

First of all, an aspect of antibodies that we did not discuss yet is that they don't just work against pathogens but also against their toxins. In the small world, a toxic substance is nothing more than a molecule that disrupts natural processes or causes damage by destroying or dissolving structures. Antibodies can neutralize these molecules by binding to them with their pincers and rendering them harmless.

So when a venomous snake bites you it directly injects you with a huge amount of harmful molecules. If we assume this is not a deadly snake that would kill you quickly, the immune processes we have learned about would be triggered. The damage and the death among civilian cells caused by the venom triggers inflammation and activates Dendritic Cells, which will ultimately lead to B Cells producing protective antibodies against this specific venom.

Think about how cool that is—the immune system is so powerful that it can produce answers to the most dangerous venoms in nature. Although in reality the bites of venomous animals are so dangerous because the damage their toxins cause is pretty much instant and only gets worse. In many cases waiting for a week until the adaptive immune system has done its thing is not an option because death will stop this process before it is finished.

So to basically cheat the system humans began producing *antivenoms*— which are nothing more than purified antibodies against the venom molecules that can be injected into the system of a person that got bitten!

The way these antibodies are made is pretty curious—venom is first harvested from a snake and then injected into mammals, like horses or rabbits, in a dose they can handle without dying. The dose is slowly increased over time so they have a chance to develop immunity against it—which means that they produce a large number of specific antibodies against the venom that saturate their blood and make them immune. This blood is then harvested and the antibodies are filtered from all the other animal blood components. Voila, you have an antivenom ready to be injected into a human that got bitten. As you may imagine this process is not entirely without risk—the human immune system can still react if there are too many animal proteins left. But usually the risk of an adverse reaction to the antivenom is dwarfed by the risk and damage from the venom itself and so it is usually administered if possible.*

Passive immunization also occurs naturally during pregnancy, when certain antibodies can pass through the placenta and enter a fetus to give it the protection of its mother.

What is even more interesting is that once a baby is born, large quantities of antibodies are passed on via human breast milk.

The process of harvesting antibodies can also straight up be done artificially from human to human—for example, in a therapy called IGIV, for "ImmunoGlobulin IntraVascular" administration, Antibodies are collected from donated blood at blood banks, pooled together, and carefully infused into patients that suffer from immune disorders and are unable to produce Antibodies themselves.

The annoying thing about passive immunization is that it is temporary. If you administer Antibodies to someone, they will stay protected as long as the Antibodies are around. But this protective effect goes away as the Antibodies are either used up or decay through natural processes. So as great as passive immunization is, it is not the best way to create immunity for most people.

It is the equivalent of giving a hungry man a fish, instead of teaching him

* Are you ready for something really cool that is also really bad? With everything that we have learned about proteins and antigens and all that, how is it possible that your immune system is cool with getting antibodies from an entirely different species? Well fun fact, it is not cool with that but actually really indignant about the sudden flood of horse or bunny protein. So while the antivenom will work fine the first time, the second time you might be immune to it because your body might make antibodies against the antibodies from a horse or bunny. This is one of the cases where the immune system just could not have expected that modern medicine would come up with creative solutions like pumping venom into a horse and then using its blood for ourselves. Which is fair so we can't really be annoyed too much with our immune system in that case.

how to fish. To really actively create immunity in people, we need to stimulate the immune system to create immunity by itself!

Active Immunization—Learning how to Fish

If you have read this far in the book you already know what active immunization is doing in your body: It creates Memory Cells that keep weapons against a specific pathogen ready.

Natural active immunization is what we explained in this book so far—for example, you get influenza A and you become immune against that specific strain forever. But this natural way has a bunch of downsides, mainly that you need to suffer through the disease to become immune to it. So the solution seems easy, all we need to do is to trick the body into thinking it is sick to become immune against all sorts of diseases!

But, of course, it is easier said than done. Because your immune system is very cautious and requires very specific signals to boot up and activate properly. To artificially induce the creation of Memory Cells that stick around for years, we need to activate your immune system for real. And this means it has to go through the proper steps of escalation. Two-factor authentication and all of that jazz.

So we must somehow *safely* provoke a proper immune response but avoid causing the actual disease we want to protect ourselves against. There are a few different ways to do this.

The first method basically comes back to the original principle of variolation. What if we could *sort of* cause the disease we want to immunize against, but just a really, really weak version of it? This is the principle of *live-attenuated vaccines*, where we put the real thing into our bodies, but a weak version of it.

The original pathogen, like a chicken pox, measles, or mumps virus, is artificially made into a pathetic shadow of itself in a laboratory. This works especially well with viruses because, in contrast to pathogens like bacteria, they are just really simple creatures with only a handful of genes, making it easier to control how well they work. The mechanism of how live viruses are weakened is pretty interesting because it is literally using evolution. A little bit how the ancestors of dogs used to be majestic and powerful wolves that we turned into pugs and Italian greyhounds.

In the case of the measles virus, for example, the virus that we use for vaccines today was isolated from a kid in the 1950s. It was cultured over and over and over again in tissue samples in a lab until it was tamed. The measles virus that had been domesticated this way is only a shadow of its former self: Weak and harmless and a pathetic version of its wild distant cousin. It still can grow and multiply but it is unable to cause a proper measles outbreak, while it still is provoking the same strong immune response as a real and dangerous measles infection would.

It can cause very mild symptoms, like a bit of fever for example, or in rare cases a sort of very mild version of a rash, again similar to the variolation experiments hundreds of years ago. One to two rounds of this vaccine are enough to create enough memory cells in kids to protect them for the rest of their lives!

Live vaccines come with downsides of course—for example, they have to be stored at the proper temperatures so the weak pathogens don't die before they can be administered. And they can't be used on people that are severely immunocompromised, as they do not have the tools to fight even weak infections. For the vast majority of people this sort of vaccination is a safe and effective way to artificially upgrade their immune system and protect them against the target of the vaccine for the remainder of their lives.

Using living pathogens is not always an option though. Just like you can't domesticate great white sharks, some pathogens refuse to be tamed and become properly weak. In some cases the risk of them causing the disease we want to protect against is just too high. So another method is to straight up kill the pathogen before injecting it, which is called an *inactivated vaccine*.

You get a high number of the pathogenic bacteria or viruses together and then you destroy them with chemicals, heat, or even radiation. The goal is to destroy their genetic code so they are empty and dead husks, unable to reproduce and go through their life cycles. But this also creates a problem. Can you imagine what?

They are *too* harmless now! Your immune system will not be properly provoked by a bunch of carcasses of pathogens that just float around and are pretty dead. So the dead remains of the pathogens have to be mixed with chemicals that highly activate the immune system. You can imagine these chemicals like insults that provoke a disproportionate response, like running around wearing the jersey of the opposing team and insulting the home team in a city that just has lost a major sporting event. Chances are at some point someone will punch you in the face.

If dead bacteria are mixed with substances that really put the immune system on edge, your immune cells will not be able to make the distinction but order the creation of memory cells. Unfortunately a number of people who don't understand chemistry have deducted from this that vaccines are filled with poison, which could not be further from the truth. For one, the doses of these chemicals are laughably small and usually only able to create a local reaction. And without them, the vaccine would not work. Another upside for this kind of vaccine is that it is much more stable and easy to store and transport than live vaccines!

Going one step further than just killing a pathogen are *subunit vaccines.* Instead of injecting a whole pathogen, only subunits, or in other words, certain parts (antigens) of the pathogen, are used so they can more easily be recognized by T and B Cells—which is a very secure way of vaccinating, as it massively decreases the likelihood of an adverse reaction to the pathogen. (This is because sometimes it is not the pathogen directly who causes harm but their metabolic products, which is a nice way of saying "bacteria poop.")

The process of making these subunits is highly interesting as it includes a tiny bit of simple genetic engineering. In the case of the hepatitis B vaccine, parts of the virus DNA are implanted into a yeast cell. The yeast cell then produces massive amounts of virus antigen that they display on their outside where it can be harvested. This way we can create highly specific parts of a pathogen and point the immune system towards it with a lot of precision! Just like other inactivated vaccines, the antigens need to be mixed with insulting chemicals that make your immune system think they are dangerous.

And lastly, let us mention the newest type of vaccine, the *mRNA vaccines.* The basic principle here is pretty genius, it is basically making our own cells produce antigens that the immune system can then pick up. Remember mRNA, the molecule that tells the protein production facilities in your cells what proteins to make? Basically, you inject someone with mRNA that will make a few of your cells make viral antigens, which the cell then showcases to the immune system. The immune system is pretty alarmed by this and will create defenses against this antigen.

There are more subtypes of vaccines but for this book these are enough details. Despite the fact that vaccines protect us from some of the worst diseases humanity has suffered from, more and more people have stopped vaccinating their children.

The reasons for the anti-vaccine movements mistrusting vaccines are diverse, but in the United States and Europe the belief that the risks of vaccines outweigh their benefits is especially prevalent. That vaccines are an artificial intervention into natural processes and that it is less dangerous to let nature run its course.

If you understand the mechanisms of the immune system and how immunity is created, this idea quickly loses all power because vaccines and diseases both do the same thing: They create Memory Cells by triggering an immune reaction. But while pathogens do this by attacking the body and causing a huge amount of stress, which carries the very real risks of various long-term consequences including death, vaccines arrive at the same goal, without all the risks of diseases.

Let us think about this in a different way. Imagine you wanted to send your kids into a dojo where they can learn some self-defense so they are prepared if someone wants to rob them. In your city there are two dojos so you arrange to visit both of them and look at the training methods. The first is "Nature Dojo." The philosophy of the head trainer is that kids should train with real weapons, real knives and swords, so they are better prepared for the real dangers of the world. After all, it is more natural and real life just is dangerous. From time to time a student will get a deep cut and require stitches. And yeah, OK, there may be a lost eye and sometimes a kid may die. But this is the natural way!

The second Dojo is called "Vaccine Dojo," and here the curriculum and exercises are basically the same as in "Nature Dojo," with one major difference. The kids use weapons made from foam and paper. Are there injuries? Well yes, very seldomly, but much, much less frequently and usually they are small bruises not even worth any tears. Which dojo would you choose for your kids, if you had to take one of them?

Let's be real, nothing in life is entirely without risk—but we can make educated decisions that are less risky and more likely to avoid damage. And in the case of vaccines, if you don't make a decision your kids are automatically enrolled in "Nature Dojo."

On top of everything else, vaccination is a sort of social contract that has benefits for all of us. If everyone that is healthy enough gets a vaccine for a disease, we are creating *herd immunity*, and are protecting everyone who is not able to. There are a number of reasons why some people can't get vaccinated. Maybe they are too young, maybe they suffer from an immunodeficiency that makes them unable to create memory cells, maybe they are currently being treated for cancer and the chemotherapy just destroyed their immune system.

Only the collective can protect these people from the diseases we vaccinate against. Herd immunity basically means that we immunize enough people against a disease so that it can't spread and dies before it reaches its victims. The problem is, for that to work we need to actually vaccinate enough people—for measles, for example, 95% of the people need to be vaccinated to create efficient herd immunity.

OK, we basically covered the most important parts of your immune system! You got to know your soldier cells, your intelligence networks, special organs, protein armies, specialized superweapons, and the mechanisms of how they work together! Now that this stuff is covered we have an opportunity to see what happens when all of these great systems fall apart. What happens when a pathogen interferes with your T Cells, what if your immune cells are fighting way too hard and begin hurting you from the inside, what can you do to boost your immune system, and how is it protecting you against cancer?

Part 4

Rebellion and Civil War

37 When Your Immune System Is Too Weak: HIV and AIDS

THE *HUMAN IMMUNODEFICIENCY VIRUS*, OR *HIV*, IS A REALLY TERRIFYING but fascinating example to showcase what happens when your immune system breaks down. Actually, it's not even your whole immune system, just a very particular cell. The main victims of the virus are your Helper T Cells. Yes, all the horror of HIV and AIDS is because it knocks your Helper T Cells out. If you read this far you probably understand how important this cell is and how much of your defense relies on them.

As a species we are incredibly lucky that HIV is not super easy to contract. It doesn't float through the air or live on surfaces but requires bodily fluids like blood or intense contact like sexual intercourse. Most HIV infections happen through sexual contact, through small unnoticeable injuries, where the virus passes through the defensive layers of epithelial cells.

HIV enters your cells via specific receptors called "CD4" that are found on the surfaces of Helper T Cells and to a lesser extent Macrophages and Dendritic Cells. HIV is a **retrovirus** which means that it intrudes and merges with your genetic code, your most intimate expression of your individuality. In a sense HIV becomes a part of you forever. But a corrupt version of you.

The human genome project found the genetic remains, living fossils, of thousands of viruses in our DNA, comprising up to 8% of our genetic code. So in a sense you are 8% virus. Most of that genetic code is useless and probably doesn't harm us. But it shows that when a retrovirus infects you, it is here to stay.

Do you remember our virus metaphor with the silent soldiers that kill citizens in their sleep? HIV is like a soldier killing their victim but then flaying their corpse and wearing the skin as a costume to walk around the city during the daytime.

HIV infections move forward in three stages:

The first stage is the *acute phase*. It is thought that Dendritic Cells are among the first cells that HIV infects and takes over. Which is great for the virus, because the Dendritic Cell will do its job, which means carry HIV to the place in your body where the cells hang out that it is looking for: The T Cell dating areas in your lymph node megacities.

Once the infected Dendritic Cell arrives here, HIV has easy access to countless T Helper Cells. So HIV really acts like a sleeper agent wearing the skin of its victims to invade the headquarters of an enemy country.

Once the virus gains access to its favorite victims, the number of viruses explodes. Early on in the HIV infection, the virus multiplies pretty much unchecked, while the innate immune system unsuccessfully tries to slow down this process. During this phase your body reacts to HIV like it does to all viruses, using the regular mechanisms and weapons, activating the adaptive system—and this is when you first might notice the infection.

The earliest symptoms of HIV are not that well described because usually the diagnosis does not happen until weeks, months, or even years after the infection. What we know is that HIV infections begin very mildly with symptoms of a harmless cold. A general feeling of fatigue, maybe a sore throat and a low fever. Symptoms like everyone experiences a few times a year, nothing you really react to strongly. It is just not a big deal.

At some point, enough Killer T Cells and Plasma Cells will be activated and devastate the virus, killing infected cells left and right and eradicating billions of viruses. Your symptoms disappear and you might think the mild cold you have is over. For most regular virus infections, this is it. Sterilization occurs as all viruses are wiped out and Memory T and B Cells are now in place to protect you from this virus for years, if not forever. And if you are really lucky this might happen in extremely rare cases of HIV infections too.

But usually with HIV this is just the beginning.

Now the *chronic phase* of the infection begins. Most types of viruses would not survive the onslaught of the immune system—but HIV has many extraordinary methods to survive: First of all, the virus does not spread just by making many copies of itself until the cell bursts—it is much more careful and works to keep its victims alive as long as possible.

Second, it has a few extra-sneaky ways of finding new victims. In Cell-to-

Cell Spread, the virus can be transmitted from one cell directly to another one. Here, HIV makes use of an important mechanism of your immune cells: *Immunological synapses.* When immune cells interact directly to activate each other, they sort of bash their faces together and lick each other's cheeks. Which means, getting very close and touching each other with many short extensions, called pseudopodia. It looks a bit funny, like many short fingers reaching out of the cells—this is the way many immune cells check each other's receptors. And these interactions can get hijacked by HIV—it uses this close connection to jump from one cell to another.

This has many upsides. The virus doesn't need to kill a cell, which would make it burst and release urgent alarm signals that make the immune system angry and alert. It doesn't require a large number of viruses floating around outside of cells, which could be picked up and cause alarm. And it has a very high success rate to infect another victim compared to the floating-around-randomly strategy that most viruses employ. This way, HIV uses interactions between cells and jumps from infected Helper T Cells to Killer T Cells, from Dendritic Cells to T Cells, from T Cells to Macrophages.

And last but not least, HIV can hide very efficiently this way. Even if the immune system flares up and kills most of the infected cells from time to time, the virus only has to sit idle in a few cells inside a lymph node to be carried around your whole body again, always in close vicinity to all the cells it wants to get close to! This also makes it much harder for medication and therapies to get rid of HIV because it has many different avenues to spread between its target cells.

HIV also can lie dormant and do nothing in cells for long periods of time waiting for the right time to become active. When a cell does not multiply, its protein production is in a sort of slow mode, aimed only at sustaining the cell—but when a cell proliferates, these production machineries amplify thousandfolds.

So when an infected Helper T Cell begins to multiply, HIV awakens and makes thousands of new viruses within hours. This is so effective that even if Killer T Cells are around looking for it, the virus is able to produce a lot of new viruses without being caught and to infect a lot of new cells.

We spoke before about the huge challenge that microorganisms present to the Immune System because of one core ability: They can change and adapt much quicker than multicellular beings and therefore we need our Adaptive Immune System to stand a chance. What makes HIV so incredibly dangerous is that it operates on a completely different level in terms of genetic variability. The genetic code of HIV is extremely prone to copying errors—on average, every time the virus makes a copy of itself it makes an error. Which means even in a single cell there are numerous different variants of HIV.

This has three possible outcomes: 1. HIV destroys itself because it mutates in a way that disables itself or it becomes less effective. 2. The mutation does not help or harm and nothing changes. 3. The virus becomes better at avoiding the defenses of the immune system.

When you are infected, HIV can produce about ten *billion* new HIV viruses in a single day—so by pure chance, a lot of viruses will be produced that are actually better at keeping the infection going. Worse still, cells can be infected simultaneously by multiple different strains of HIV, and these strains can be recombined into new hybrids. If you can try out billions of new versions every single day, chances are really good that a few of the new viruses are great at avoiding the immune response.

Now think about what this means: It took the Adaptive Immune System about a week to make thousands of Killer T Cells and millions of antibodies that are extremely good at hunting HIV down—but already there are numerous new viruses that have new and different antigens! Different enough that the Killer Cells and Antibodies you just made may be useless against them.

And now the new and different viruses infect new cells and make millions of copies of themselves, again. For them, the virus your Adaptive Immune System adapted to is already old news and irrelevant. HIV is always a step ahead of the immune system. And so in the chronic phase of an HIV infection your body is still teeming with the virus. On average, in this phase, a single milliliter of blood contains between 1,000 and 100,000 virus particles.

So let us summarize the tactic of HIV real quick before we move on: By infecting Dendritic Cells, the virus gets a taxi into HIV heaven: The lymph nodes, which are filled top to bottom with Helper T Cells. HIV can build reservoirs in these cells and stay hidden indefinitely. When Helper T Cells begin to proliferate massively, they do so at lymph nodes, which is the ideal place for HIV to also make millions of new viruses. So the place that is most central to building protection against viruses is completely taken over and actually becomes a weak point.

This is still not the worst part. Think about what HIV really does by specifically attacking T Cells: It destroys and kills the cells that the Adaptive Immune System needs to properly activate B Cells and Killer T Cells.

Still your Immune System does not give up. An epic struggle begins that will go on for years. Every day, HIV makes billions of new viruses and your immune system reacts in kind with new antibodies and new Killer T Cells. A push and pull of death and rebirth, a struggle for survival on both sides. This struggle can take up to ten or more years and usually comes with very few noticeable side effects—which in a perverse twist enables an infected person to serve as a reservoir and infect others.

And although your immune system is giving its all, the cards are stacked against you. Not only are your Helper T Cells constantly being infected by HIV, Killer T Cells viciously hunt them down too. (That is because, if your Helper T Cells show HIV antigens in their display windows, Killer T Cells will order them to kill themselves!) Which is good in principle but also means that the weapons you need against HIV are getting exhausted.

But not just the Helper T Cells, Dendritic Cells suffer too—and they are just as important to activate the immune system. Without these two cells, the ability of the Adaptive Immune System to mobilize begins to break down. This dying goes on for years as the body desperately produces new Helper T Cells—but over the long term it just can't keep up. As the years pass, the total amount of Helper T Cells is slowly getting lower and lower and lower. Until one day a critical threshold is reached and the Adaptive Immune System collapses. The amount of virus particles in the blood explodes and saturates the body as there is only extremely little resistance left.

The last stage begins: *profound immunosuppression*. AIDS, the *Acquired Immune Deficiency Syndrome,* begins. What this means is that your Adaptive Immune System is basically out of order, which demonstrates how tremendously important it is. Hundreds of pathogens, microorganisms, and cancers that are usually not the slightest problem for your body now quickly become dangerous and lethal. But not only are you now extremely susceptible to countless diseases from the outside. Because to fight cancer, you need your Adaptive Immune System, Helper and Killer T Cells in particular, cancer now can thrive with very little opposition. If AIDS breaks out the situation quickly becomes dire and dangerous. The leading causes of death are various forms of cancer and bacterial or viral infections, often a combination of all three. Basically everything your immune system usually protects you against.

HIV infections used to be a death sentence, with the disease marching towards an eventual outbreak of AIDS that was soon followed by death. But thanks to an immense and unparalleled effort of the scientific and medical community, for people receiving proper treatment, HIV has turned into a chronic disease that is manageable. Almost all therapies for HIV are targeted at preventing the last stage—to prevent AIDS from ever breaking out, because this is where people die.*

---

* A natural question at this point is how HIV treatments work. Well, without going into too much detail, the mechanisms are more or less that the different stages of virus development are targeted and blocked or slowed down, so the HIV infection can never turn into AIDS. The more interesting question, though, is why don't we have working medicine against the flu but we do have a few different treatments against HIV? (Well OK, we actually do have a very safe and effective vaccine against the flu, which is redeveloped each year to account for the rapid mutation of the flu virus. It is just that for some reason not that many people get flu shots.) OK. The answer is a bit depressing: Attention and money. It is easy to forget that HIV used to be a new pandemic once and a very shocking and scary one. In 2019 there were still around thirty-eight million infected people around the world. When HIV and AIDS emerged it caused panic in the establishment, leading to an outpouring of resources and attention that was unprecedented. Humanity really wanted to get results, fast (as a happy byproduct, immunologists learned a lot of new things about the immune system). And we got them, and turned HIV from deadly to a chronic disease and may even slay it for good one day. Similar things could be observed with the vaccines for COVID-19, which broke even the best speed record. In the end it really seems to be a question of what is a cure worth to us and how desperately do we want it. Another testament to the fact that humans really could solve all of their major problems if they were better at prioritizing.

38　When the Immune System Is Too Aggressive: Allergies

YOUR WHOLE LIFE CRABS WERE AMONG YOUR FAVORITE FOODS, THESE funny-looking giant spiders crawling around on the bottom of the ocean, with their weird texture and great taste. After being good and dieting for a few months, today was supposed to be a night of indulgence with your friends, wine, and a lot of crabs. But soon after your first bite something weird happened. Your body began feeling funny, a little bit on edge.

You felt hot and began to sweat, your ears, face, and hands felt weird, and suddenly you noticed that breathing got harder and you panicked a bit. Your friends asked if you were OK as you got up but had to sit down immediately because you felt so dizzy. And then you woke up in an ambulance racing to the hospital, a needle in your arm dripping chemicals into your system, which calmed down the allergic reaction that almost killed you. You are confused but also relieved to be in the care of professionals, when a sudden realization hits you: You will never be able to eat crabs again.*

As we have seen numerous times in the book, the immune system walks a very tight line. If it doesn't react strongly enough even smaller infections could turn into deadly diseases quickly and kill you. But if it reacts too strongly then it can do more damage than any infection—your immune system is much more dangerous to your survival than any pathogen ever could be. Think of Ebola, even this pretty disgusting and horrible disease needs about six days to kill you. Your immune system has the power to kill you in

---

* It is very likely that a few people reading these pages had an experience at least somewhat similar. Quite a few more will have unpleasant but not directly life-threatening experiences. Shellfish is the most common food allergy that adults can suddenly gain but there are a lot of different things that you can become suddenly allergic to, from milk to nuts, soy, sesame, eggs, or wheat. Allergies suck.

about fifteen minutes. People who suffer from *allergies* have experience with this dark side of their defense network. When your immune system loses its self-composure, it becomes deadly, killing a few thousand people by anaphylactic shock each day. Why would your immune system do something like that?

Being allergic means that the immune system massively overreacts to something that might not be all that dangerous. It means that it mobilizes forces and prepares to fight, although there is no real threat present. About one in five people in the West suffer from some form of allergy—most commonly *immediate hypersensitivity*, which means that symptoms are triggered very quickly, within minutes of being exposed. It is a little bit like finding a bug in your living room and calling the military to wipe out your city with tactical nuclear weapons. Sure—this deals with the bug but maybe it is not necessary to turn your house into a glowing pile of rubble to do that. The most common immediate hypersensitivity reactions in the developed world are hay fever, asthma, and food allergies, with various degrees of seriousness. You can be allergic to basically everything.

Some people are allergic to latex and can wear neither latex gloves nor full-body latex suits (which is a real tragedy if they are into that). Others are allergic to the stings from certain insects, from bees to ticks. There is a diverse set of food allergies and of course you can be allergic to any kind of drug.

What your immune system is reacting to are antigens, the molecules of harmless substances. In the context of allergies, antigens are called *allergens*, although they are functionally the same: A short piece of protein, say from crabmeat, that can be recognized by your adaptive immune cells and antibodies and that causes allergy is an allergen.

Why does your immune system think any of this is a good idea? Well, it doesn't. It doesn't think or do anything with purpose, there are just mechanisms that misfire horribly. In this case, the source of your immediate hypersensitive reaction is in your blood. Here the most annoying part of your whole immune system does its thing: The *IgE Antibody*. You can thank the IgE Antibody for a lot of your allergy-related suffering. (They actually have an important job they don't get to do as much nowadays, but more on that in the next chapter.)

IgE is produced by specialized B Cells that tend not to be stationed in your lymph nodes, but in your skin, lungs, and intestines: Where they can do the most damage—presumably to enemies that might overcome the walls of your defenses—but in reality mostly to you. What do IgE Antibodies actually do when you suffer from an allergic reaction?

Hypersensitivity always happens in two steps: First you need to encounter your new mortal enemy. And then you have to meet them again.

Say, for example, you eat a meal like crabs or peanuts or get stung by a bee. The first time, all is good. The allergen floods your system and for some reason, B Cells that are able to bind to them with their receptors are activated. They begin making IgE Antibodies against the allergen, like for example, crabmeat proteins, but for now things are chill, nothing happens. You can imagine this step as arming the bomb. (In cases like our poor protagonist at the beginning of this chapter it is unclear when and why exactly this arming occurred—but it has to occur at some point.)*

Now, after the exposure to the crabmeat, a lot of IgE Antibodies that are able to attach to the crabmeat allergen are in your system. But IgE Antibodies by themselves would not be problematic as they are not particularly long-lived and dissolve after a few days. They need help to become a problem from a special cell in your skin, lungs, and intestines that is especially receptive to IgE Antibodies: the *Mast Cell*.

We met it briefly before when we talked about inflammation. To refresh your memory, Mast Cells are large, bloated monsters filled with tiny bombs that carry extremely potent chemicals like *histamine* that cause rapid and massive inflammation. The job of the Mast Cell is still debated by scientists—some think it is crucial to early immune defenses and others give it more of

---

* Urgh. OK, time for a big BUT. So what we are describing here is the "standard" case of how allergies work. You encounter an allergen for the first time, your immune system charges up, you encounter it a second time, and boom, allergic reaction. But what about stories like our introduction, with the poor person suddenly no longer able to enjoy their favorite ocean spider dish? Well, here is a fun fact: We don't exactly understand this yet. Adult-onset allergies are a bit of a mystery, which is a bit frightening considering how many people encounter them in their lives. I myself had the joy of being rushed to a hospital with a new surprise allergy to something I had eaten for years, so I'd very much like to know how this works. But yeah, you now have to live with the information that people can just suddenly become allergic to things they ate all their lives with no warning.

a secondary role. What we know for sure is that Mast Cells serve as inflammation superchargers. And unfortunately, they do their job with a bit too much enthusiasm in the case of allergic reactions.

Mast Cells have receptors that connect and attach to the butt regions of IgE Antibodies. So if IgE is produced after your first exposure to an allergen, Mast Cells swoop them up like a large magnet would swoop up a bunch of nails. So you can imagine a "charged and armed" Mast Cell as a big magnet, covered with thousands of tiny spikes. When allergens pass by, the IgE Antibodies on the Mast Cell can extremely easily connect to them. To make matters worse, IgE on Mast Cells is stable for weeks or even months—the connection protects them from decay. So, after your initial exposure to an allergen, you have these bombs in your skin, your lungs, or your gut, ready to get activated very quickly. Time passes and nothing happens, until you finally eat a bunch of crabmeat again and flood your system with the allergen enabling the Mast Cells covered in IgE to connect to it. Now the armed allergy bomb goes off inside your body.

Your armed Mast Cells undergo degranulation, which is a nice way of saying that they vomit out all of their chemicals that are inflammatory superchargers, especially histamine. This causes virtually all of the very unpleasant things that you experience during an allergic reaction: It tells the blood vessels to contract and let fluid stream into the tissue, causing redness, heat and swelling, itching, and a general feeling of unwellness.

If this happens in too many regions of the body at the same time, it can lead to a dangerous loss of blood pressure, which can be deadly on its own. Histamine also stimulates the cells that produce and secrete mucus to step up their game, giving you an extra, unnecessary flow of snot and slime in your respiratory system.

But most dangerously, histamine can cause the smooth muscles in your lung to contract and make breathing hard or even impossible. It is not so much that you can't get air in, it is more that the air inside your lung is trapped and getting it out again becomes really hard. All the extra slime your mucous membranes produce is definitely not helping this situation. As there are a lot of Mast Cells in the lungs, allergic reactions that happen here can become dangerous very quickly, as extra fluid and mucus fill up the lung, while it becomes harder and harder to actually breathe. The worst case

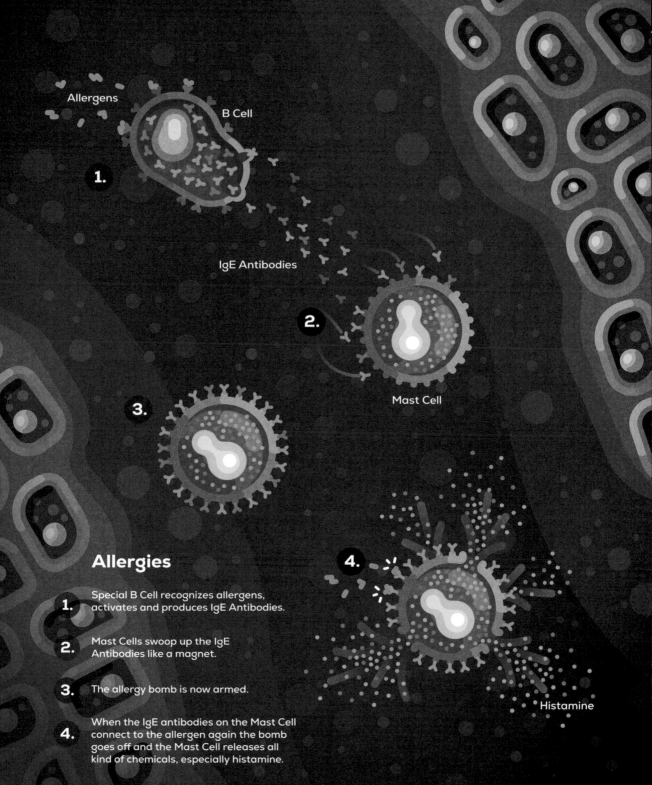

Allergens

B Cell

1.

IgE Antibodies

2.

Mast Cell

3.

4.

Allergies

1. Special B Cell recognizes allergens, activates and produces IgE Antibodies.

2. Mast Cells swoop up the IgE Antibodies like a magnet.

3. The allergy bomb is now armed.

4. When the IgE antibodies on the Mast Cell connect to the allergen again the bomb goes off and the Mast Cell releases all kind of chemicals, especially histamine.

Histamine

can be an anaphylactic shock, which can kill within a few minutes. Allergic reactions are no joke.

We have given the Mast Cell a bad rap in the last few paragraphs but this is slightly unfair. Because it does not cause all of this mess alone, it has an equally harmful buddy. Once Mast Cells activate and degranulate, they also release cytokines that call for allergy reinforcements by another special cell:

The **Basophil**. Basophils patrol the body in the blood until they get called in. They also have receptors for IgE that they charge up after the initial exposure to the antigen. Basophils serve as a sort of second wave of horribleness. Once Mast Cells have caused the first wave of allergic reactions, they need to replenish their destructive histamine bombs and are temporarily out of order. Basophils fill that gap and make sure that the allergic reaction doesn't stop too quickly. They are probably really proud of themselves too, thinking that they are doing important work, innocently setting the body on fire while you are scratching your skin or emptying your inflamed bowels. These two cells are responsible for the immediate hypersensitivity.

But unfortunately this is still not the end of it. As many sufferers of asthma sorrowfully know, some allergic reactions are more of a chronic occurrence than a onetime event that ends. Let us meet the third (and luckily last) cell that thinks allergic reactions are such a great idea!

The **Eosinophil** makes sure that the symptoms of an allergic reaction stay around for a while—only a few of them exist inside your body and they tend to hang out in the bone marrow, far away from the action. Cytokines released by Mast Cells and Basophils activate them but they take their sweet time, proliferating and cloning themselves for a while before they arrive late to the party, where they unfortunately repeat the mistakes made before and cause inflammation and misery. You may rightfully ask now: Why do your own immune cells do this?

The truth is we don't know yet why some people produce a lot of IgE Antibodies when they come in contact with certain allergens and others don't. But while we don't know for sure why some people are more affected than others, we think we know what IgE Antibodies were originally supposed to do:

They are the immune system's superweapons against large parasites that are too big for your phagocytes, your Macrophages, and your Neutrophils to swallow. Especially one of the most horrible parasites: *Parasitic worms*. A menace humanity has had to deal with for millions of years. Let us learn about the true purpose of IgE Antibodies and at least somewhat clear their bad name a bit.

39 Parasites and How Your Immune System Might Miss Them

PARASITES COULD PROVIDE SOME ANSWERS TO THE ANNOYING NATURE OF allergies. One of the worst things to do late at night is to google infections by parasitic worms. You can ruin your life even more if you click on image search. Of all the possible pathogens and parasites that can victimize humans, worms are by far the most upsetting ones. Nothing quite compares to a faceless, slimy, stringy thing that drills itself through your insides, pooping, laying eggs, spending its whole life inside you. It's right out of a horror movie.

There are nearly 300 species of parasitic worms who can infest humans. And while only around a dozen of these species are widespread, they still infect up to two billion people, close to a third of humanity. Most species of these parasitic worms tend to establish stable chronic infections that can persist for up to twenty years, while their eggs or larvae leave your body with your poop. Parasitic worms thrive in underdeveloped rural regions or in slums, where unsanitary conditions and dirty water combine into an environment perfect for parasites that leave the body on one end and enter through another.*

Being infested by parasitic worms is not a great experience. Take hookworms, for example, parasites that are about half an inch long and live in

---

* Unfortunately, as parasitic worm infestations correlate with poverty and low infrastructure development, there is another bonus problem on top. If you suffer from malnutrition, a parasitic worm is a much larger problem for you than if you are well fed. Which makes sense because on a fundamental level, the worm is inside you because it wants to steal your nutrients. If you have a hard time getting enough calories for yourself, having subtenants in your body that don't pay rent can seriously weaken your whole system. So the people who are the least fortunate suffer the most from these parasites.

your guts. Their name is their game, as they hook themselves into the walls of your intestines and can cause extensive blood loss. This in turn can cause anemia, the lack of healthy red blood cells to carry enough oxygen to your organs and tissues, weakening the whole body. Infested people have greenish-yellow pallor and are tired, weak, and generally low on energy. The hookworms produce eggs that leave through your feces and when the eggs turn into larvae, they bore through the skin of a new host and migrate to the lungs from where they ultimately end up in the small intestines again, to repeat the cycle anew.

Seriously, thanks, but no thanks.

Parasitic worms are genuinely not fun at all. Until pretty recently in human history, infections by worms were widespread and basically unavoidable.*

Against parasitic worms, the weird mechanism of the IgE Antibody suddenly makes a lot of sense. From the scale of an immune cell, worms are giant monsters, reaching far into the sky beyond the horizon.

Attacking them needs to come with some punch to even have some hope to do damage. It requires a lot of combined effort from the immune system to kill a worm and rid the body of its presence. Millions of years ago, the immune system of our ancestors came up with a strategy: Stage one is to recognize the worm and prepare a brutal attack.

So when the worm is recognized for the first time—probably close to the border regions of the body—the special B Cells stationed near the skin or in the respiratory or intestinal tracts begin the preparation by making large amounts of IgE Antibodies. These IgE Antibodies "prime" your Mast Cells—if you consider Mast Cells weapons, IgE Antibodies activate them and remove the safety pins. If the immune system then encounters the worm again, Mast Cells can connect to them with the IgE Antibodies on their surfaces and vomit their harsh weapons directly onto them from very close range. Not only does the mix of chemicals damage and hurt the worm, but the harsh and immediate inflammation that the Mast Cell triggers alerts the rest of the immune system. Macrophages and Neutrophils will swarm in and continue to attack the worm. The Basophils will be alerted by the com-

---

* Well, actually, they are still widespread. Just not in developed countries.

motion making sure the attack will not be exhausted before the worm is actually killed. The Eosinophils from the bone marrow move in later and continue to attack the worm and its eventual buddies in the coming hours and days.

With this combined effort from these different cells, parasites like worms can be killed by your immune system. This is a good moment to marvel again about the huge variety of dangers our ancestors had to deal with—and how their immune system found ways to handle all of them. But we were talking about allergies originally, so let us find the connection to our horrifying wormy enemies.

As you may imagine, parasitic worms are not happy about IgE and Mast Cells and being attacked and all—since they are living beings specializing in . . . well . . . being parasites, they evolved to deal with our defenses whenever possible. In this case this means shutting down your defense. Parasitic worms that have adapted to humans are able to modify and recalibrate almost every facet of their host's immune system. They employ a wide range of immunosuppressive mechanisms. Or put simply: Worms release a plethora of chemicals to downregulate and modulate your immune system to make it weaker.

This has a large variety of consequences, some intended and some unintended. For one, a weaker immune system is worse at preventing infection from viruses and bacteria and it might have a harder time catching cancerous cells before they become a deadly threat. But not all of the effects are actually bad. Worms suppress the mechanisms that cause inflammation reactions, allergies, and autoimmune diseases.

We will learn a bit more about autoimmune diseases in the next chapter—but in a nutshell, if the immune system is downregulated to be less aggressive, it also can't do as much damage to the body. Because of this fact, some scientists argue the lack of worms in humans in the developed world is weird for your immune system because it evolved assuming that you would suffer from parasitic worms on a regular basis.

Our ancestors were basically helpless when it came to parasitic worms. They did not have any drugs against them, they did not understand the nature of hygiene, and they often didn't have access to clean drinking water in the environment they lived in. So their bodies, begrudgingly, had to adapt to frequent, if not permanent infestations by parasitic worms. One of these adaptations might have been to upregulate the aggressiveness of the immune system. Basically making it a bit too aggressive, so that in spite of the suppressing effects of the worms, it was strong enough to deal with infections and infestations by pathogens. A sort of deal with the devil that our immune systems had to enter into millions of years ago.

In evolutionary terms, humans in developed countries in the last few hundred years or so suddenly lost their parasitic wormy guests. The advent of soap and hygiene and the clear separation of poop and drinking water destroyed the life cycles of most of the worms that were living within us. The remaining worms were exiled by drugs and modern medicine.

Which left our immune system suddenly without the enemy that had kept it down a notch for millions of years. And so it might be that our immune system still operates under the assumption that worms are making it weaker and that it needs to be more aggressive as a counterbalance.

If this general idea is true it might explain a lot of diseases caused by too-aggressive immune systems in people without worms, mostly allergies and inflammatory disease. And not only that—the lack of worms leaves a whole bunch of our cells without the enemy they were made to fight on a regular basis. So the idea makes sense that without wormy stimulation, these weapons just found new targets.

But while parasitic worms may be a piece of the puzzle, they alone are not nearly enough to explain the rise in the prevalence of allergies and the rise of a set of much more serious disorders that affect millions of people: *Autoimmune diseases*. They are what happens when the immune system thinks your body is *other*—and that it needs to be destroyed.

40 Autoimmune Disease

Your body takes autoimmunity very seriously, as we learned by looking at the Murder University of the Thymus, where only cells who could distinguish *self* from *other* were allowed to live. And it became apparent with the many hoops your T Cells and B Cells need to jump through before they can get activated and actually start doing their job. But still, despite all the security systems and different layers that are supposed to prevent your immune system from attacking your own body, things can go terribly wrong. Security mechanisms can fail in such an unlucky sequence of events that your immune system thinks the body it was made to protect is the enemy that it needs to kill.

It would be as if the army of a nation suddenly pointed its weapons at its own defenseless cities and infrastructure. Tearing up roads, bombing civilian centers, shooting at the construction workers, baristas, and doctors who are just trying to keep society running. It would be all the worse because if the military attacks its own country and is really committed—who, exactly, could stop it? In a way this is what autoimmune diseases are. While civilian cells try to keep everything together, to get resources to everyone, and to keep the bodily infrastructure and organs intact, parts of the army tear it down again and shoot civilians in the head.

Autoimmune diseases don't just happen though. For most people they are a colossal case of bad luck. Although obviously things are a bit more complicated in reality, we can look at the basic principles. In a nutshell, in autoimmunity, your T and B Cells are able to recognize proteins that are used by your own cells. *Self-antigens.* The antigens of *self. You.*

This might be a protein on the surface of a liver cell, an important molecule that keeps you alive like insulin, or a structure that is part of a nerve cell, for example. If misguided T and B Cells connect to these self-antigens, your adaptive immune system mounts an immune response against your own body. So parts of your immune system are no longer able to distinguish correctly between *self* and *other—they think that self IS other*. This is various degrees of bad—from annoying, to quality-of-life ruining, to deadly.

What needs to go wrong for your immune system to get so horribly confused? Well, there are a few stages, a few conditions that need to be met:

First of all, your MHC molecules actually need to be physically able to bind to your own self-antigen efficiently. This is mostly genetic, and as everything that is etched into our genetic code, bad luck. You can't choose your parents and you can't choose your genetic makeup. (At least not yet.) In an earlier chapter we talked about the fact that MHC molecules vary a lot between different individuals and come in a few hundred slightly different shapes. Not all of these shapes are great, just by a whim of nature, some types are pretty good at presenting self-antigen. There is a hereditary risk for autoimmunity that varies between all of us—so while *everybody* can get autoimmune diseases, chances are higher for some, those with genes that make specific MHC molecule types. But just a genetic predisposition is not enough.

The second thing that needs to happen for an autoimmune disease to develop, is that you need to produce either a T or a B Cell that is actually able to recognize self-antigen and that does not get killed by your own body. Each day you make billions of T Cells, for example, and just by pure chance millions will come with receptors that can recognize self-antigen efficiently. Most of these cells do not survive their training in the thymus or bone marrow, but sometimes these mechanisms fail and they get released into circulation. Chances are that right now, you have some T or B Cells inside you that could cause an autoimmune disease. But their presence alone is still not enough, they need to be activated.

And here it becomes very tricky. We spent a good portion of this book talking about the fact that your Adaptive Immune System does not just activate by itself. It needs the Innate Immune System to make the decision to activate it, and for that you first need a battleground. An environment that can push your Innate Immune Cells to escalate an immune reaction. How exactly these things happen is hard to say and even harder to observe in living humans—people get sick all the time but only extremely rarely does this lead to anything more than an infection that is cleared out eventually. But for most autoimmune diseases, these seem to be the steps that cause them:

Step one: There are individuals who have a genetic predisposition. (Which is not a required step but it greatly enhances your chances.)
Step two: They make B or T Cells that are able to recognize a self-antigen.
Step three: An infection provokes the Innate Immune System into activating these faulty B or T Cells.

How exactly would infections cause autoimmune disease though? While still not entirely answered, a popular idea among immunologists is called *molecular mimicry*. It basically means that the antigens of microorganisms can be similar in shape to the proteins of your cells, your self-antigens. Now for one, this can happen by accident. Certain shapes are just useful in the tiny world, and while there are a great variety and a lot of options for shapes, some shapes still can be similar.

But some pathogens will also try to mimic the shapes of their host, which makes a lot of sense, since this is a mechanism we can observe plenty in the animal kingdom: Camouflage is hugely beneficial if you have to survive in a world of roaming hunters. And so from butterflies that try to look like leaves to white partridges that blend in to snow and crocodiles that disappear in muddy water, a wide range of animals try to be as hard to spot as possible. For a pathogenic virus or a bacteria your tissue is a jungle full of angry predators that are looking for them, so mimicking the environment to become harder to spot is an effective strategy.

To explain it properly, let us add a bit more detail to a simplification we've made until now. When we talked about the largest library in the universe we said that each T and B Cell is made with a special receptor, to recognize *exactly one specific antigen*.

Well, it is a bit more complex. In actuality, the range of individual T and B receptors is a bit wider. Each receptor is *extremely good* at recognizing one specific antigen. But it can also connect to a few more than exactly this one.

So a B Cell receptor for example might be extremely good at recognizing

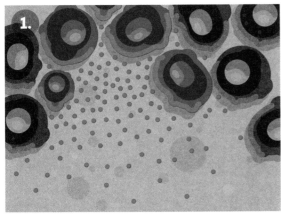

Everything begins with a pathogen infecting the body.

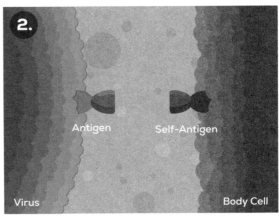

The virus has an antigen, that is similar to a self-antigen.

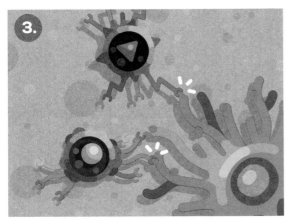

After sampling the battlefield, a Dendritic Cell activates T Cells that can connect to both the antigen AND the self-antigen.

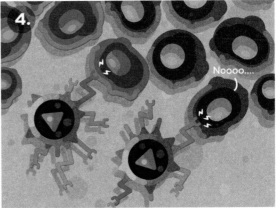

Killer T Cells start killing both infected cells and healthy cells, if they present self-antigen.

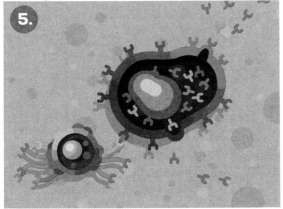

Meanwhile the Helper T Cell activates B Cells. After optimizing themselves they release **Autoantibodies** that connect to your own cells and mark them for death.

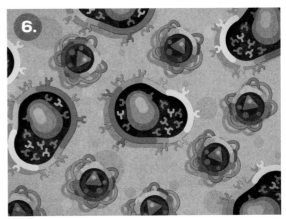

When B Cells and Killer T Cells turn into Memory Cells the autoimmune response turns into a chronic autoimmune disease.

one specific antigen and OK-ish at recognizing, say, eight different other antigens that are similar but not exactly the same.

It's kind of like when you are doing a puzzle, and you find two pieces that *almost* fit together perfectly. Like there is still some space left and they don't connect perfectly but it does hold if you don't pull too hard.

Now, let us imagine how you might get an autoimmune disease in reality. In our example everything begins with a pathogen, maybe a virus that has an antigen that is similar to a self-antigen. For example, it could be similar to a common protein that is inside your cells. When the virus enters your body and begins doing what pathogens do, civilian cells, Macrophages, and Dendritic Cells release massive amounts of cytokines and cause inflammation. This triggers Dendritic Cells to collect samples of the virus antigen, which is so similar to our self-antigen. And it triggers all the cells near the battlefield to make more MHC class I molecules, and showcase more of their internal proteins.

In the closest lymph node, your Dendritic Cell may then find a Helper T Cell or a Killer T Cell that can connect extremely well to the antigen of the enemy. *And because it is similar to a self-antigen, the T Cell receptor also is pretty OK-ish at connecting to the self-antigen that the antigen is similar to.* The Killer T Cells enter the battlefield and begin killing infected cells. But not just infected cells, they also find healthy cells that present the self-antigen, the one that is similar to the virus antigen, in their display windows. And so the Killer T Cells begin killing perfectly innocent and healthy civilian cells. Now the context of a real infection being active right now becomes crucial. Because the Killer T Cells are stimulated and activated by the real infection going on, by all the right cytokines and battle signals, some of them will turn into Memory Killer T Cells. Even after the actual infection has been wiped out, these cells will find the autoantigen (or self-antigen) presented by civilian cells and just assume there are still a lot of enemies around.

And as soon as this happens, the accidental autoimmune reaction turns into an autoimmune disease. Now the adaptive immune system thinks that it has been activated to fight the self-antigen and the body cells that express it. And why wouldn't it? In the case of Murphy's Law, every single thing that could go wrong did go wrong, and all the conditions for a proper activation have been fulfilled. It still can get worse though! Meanwhile, the activated

Helper T Cell starts activating B Cells that can accidentally fine-tune themselves on the self-antigen!

Remember, when activated B Cells begin an optimization process to refine their Antibodies, they mutate and produce a bunch of different variants, so they can become much better at fighting an enemy. But in this case, they can actually optimize themselves to the self-antigen. In the worst case, if such a B Cell gets a confirmation signal from a Helper T Cell, the immune system produces Plasma Cells that release *autoantibodies* that connect to your own cells and mark them for death.

And when B Cells mature into Plasma Cells, Memory Cells are created as a byproduct. So now all of a sudden, in your bone marrow, Long-Lived Plasma Cells begin regularly pumping out autoantibodies against your own body. They will live for years and decades. Once your adaptive immune system has made Memory Cells against your own cells, it is sure to be restimulated over and over again—as your self-antigens are, well, everywhere inside your body. These cells now find themselves in a giant world where everybody is an enemy—it's like the joke of the guy driving on the freeway when his wife calls him to warn him that she heard on the radio that there is a driver going the wrong way. And he replies in a very distressed voice: "Honey, there isn't one, there are hundreds!"

No matter how many civilian cells your immune system kills, your body will make more—and so chronic inflammation, chronic immune system activation, is the consequence. Your misguided immune cells think they are perpetually surrounded by enemies and act like it.

Although we are talking about a diverse set of different diseases, there are many common symptoms among all of them: Fatigue, rashes, itching, and other skin problems, fever, abdominal pain and a variety of digestive issues, pain and swelling in the joints. Autoimmunity is rarely fatal, it is not so much a group of diseases that kills. It is more that it makes life miserable and draining. The treatment options are somewhat limited—after all to eliminate the root of autoimmune diseases you would need to find the individual Memory Cells among billions of B and T Cells, which target your self-antigen and kill them. So at least for now, there is no cure available for autoimmunity—once you have it, you need to deal with it. To alleviate pain and inflammation, generally autoimmune diseases are treated with a variety of medication that suppresses the immune system, particularly inflammation, which as you may imagine, is not great either. It may alleviate the symptoms of autoimmunity by making the immune system weaker and less likely to attack the body but it also leaves the patient more vulnerable to infections.

An Aside **Anergy**

A SHORT SIDE NOTE THAT IS WAY TOO COOL NOT TO TELL IS ABOUT *ANERGY*, which is a passive and pretty ingenious tactic your immune system deploys to deactivate T Cells that are autoreactive, meaning able to recognize your own cells.

First let me clear up another simplification (which sounds much better than a convenient lie that made it easier to get to the point where we are now). So I talked a lot about Dendritic Cells and that they start sampling a battlefield when they get activated. Well, that is not exactly right, they actually are in a constant sampling mode all the time. Even if there is no danger, a few of your Dendritic Cells, for example in your skin, will take samples of the stuff floating around in the natural, healthy environment between your cells—a lot of it is self-antigen presumably—and then move to your lymph nodes to show the Adaptive Immune System what it found.

You may ask now: How on earth is this a good idea? Wouldn't a Dendritic Cell that collects self-antigen cause autoimmune diseases? Well, think again—what is one of the main jobs of the Innate Immune System? To provide *context* to the Adaptive Immune System. So a Dendritic Cell that moves into a lymph node with the context of "everything is fine—this is what I have to show you" can actually prevent autoimmune diseases. Because what it is really doing is "hunting" for T Cells that are *autoreactive*. That is, able to bind their MHC molecules to your self-antigen. If the Dendritic Cell finds one of these autoreactive T Cells by pure chance, it connects to it to stop it from further wrongdoing.

Remember the "kiss" signal the Dendritic Cell gives to T Cells to activate them? The confirmation signal that tells the T Cell that the danger is real?

Well, if there is no danger, the Dendritic Cell withholds this kiss signal. And a T Cell that receives an activating signal on their MHC molecules but no loving kiss on the cheek deactivates itself. It doesn't die immediately, but it can't get activated again. It is a lame duck from now on and just floats around for the rest of its lifespan before it destroys itself without causing a fuss. So, as constant background noise when you are not sick or injured, your Innate Immune System uses its free time to low-key fight autoimmune diseases. The level of overlapping systems and how all the different principles of activation and regulation work together to protect you in every possible way is just so, so fascinating. The concert of your immune system uses every available instrument to keep you safe.

OK, now that we've talked about allergies and autoimmunity, let us venture out a bit and explore why so many people seem to be affected by them.

41 The Hygiene Hypothesis and Old Friends

THE LATTER HALF OF THE TWENTIETH CENTURY SAW TWO REALLY WEIRD and counterintuitive trend lines in developed countries. While dangerous infectious diseases like smallpox, mumps, measles, and tuberculosis were successfully pushed back and in some cases got to the verge of elimination, other diseases and disorders began to grow or even skyrocketed. The rates of diseases like multiple sclerosis, hay fever, Crohn's disease, type 1 diabetes, and asthma have increased by as much as 300% in the last century. But this is not all: It seems like you can draw a direct line from how developed and rich a society is to how much of its population suffers from some kind of allergy or autoimmune disorder.

The number of new cases of type 1 diabetes is ten times higher in Finland than it is in Mexico, and 124 times higher than in Pakistan. As many as one in ten of all preschool children in Western countries suffer from some form of food allergy while only around two in one hundred in mainland China do. Ulcerative colitis, a nasty inflammatory bowel disease, is twice as prevalent in Western Europe as in Eastern Europe. Around 20% of all U.S. Americans suffer from allergies. All of these disorders have two common denominators: either the immune system is overreacting to a seemingly harmless trigger, like pollen from blooming plants, peanuts, the excrements of dust mites, or air pollution (in a nutshell: allergies), or it is going a step further and is straight up attacking and killing civilian body cells, which we experience as autoimmune disorders like type 1 diabetes. All while humans are dying less from infections.

In the late eighties, a scientist found that the rate of certain allergies was connected to the number of siblings a child had. So he asked the question if "unhygienic contact" between siblings would lead to higher rates of infections during childhood and if that could have a protective effect against allergies. And thus the *Hygiene Hypothesis* was born and almost right away it was a victim of its own appeal. The message was too straightforward, too perfect, it fit too nicely into the zeitgeist.

The perceived message was clear: In our fervor to rid ourselves from the causes of disease, humans had become too clean and sterile and had committed a sin against nature, and now we were suffering immune disorders because of it! It seemed logical that the human immune system actually needed harmful infections to function properly. The solution seemed to be equally easy and straightforward! Just be less clean, stop washing your hands, maybe eat a little spoiled food, pick your nose. In short: expose yourself and your children to microorganisms and maybe even contract more infectious diseases to train your immune system!

But as so often with the immune system, the reality seems to be way more complicated and nuanced. Today many scientists are quite upset how the Hygiene Hypothesis has permeated popular culture and thinking. Because it leads to "gut feeling" conclusions by laypeople that are, at the very least, extremely questionable if not straight-out wrong. For example, a widely held view is that it is good for us to contract diseases because surviving them makes us stronger, since this has been the natural way humans operated in the past.*

Maybe we need hostile bacteria as sparring partners to grow up strong, and maybe this mechanism of immune training has been destroyed by the modern world with all its fancy medicine and technology stuff.

---

* What is generally troubling about these appeals to naturalism is the idea itself, that something natural is somehow better. Nature does not care about you or any individual at all. Your brain and body and immune system are built on the bones of billions of your would-be ancestors who were not fast enough to escape a lion, were killed by a mild infection, or were just a little worse at pulling the nutrients from their food. Nature gave us charming diseases like smallpox, cancer, rabies, and parasitic worms that feast on the eyes of your children. Nature is cruel and without any form of care for you. Our ancestors fought tooth and nail to build a different world for themselves, a world without all this suffering and pain and horror. And consequently we should celebrate and marvel at the enormous progress we've made as a species. While we obviously still have a long way to go and the modern world has a lot of downsides, the notion that "natural is better" is something only people who are not actually living in nature can say, and who have forgotten why our ancestors worked so hard to escape it.

Discussing this topic is a tiny bit sensitive because the scientific community has not yet reached consensus and there is still a lot that we don't know or understand about the microbiota around us, our personal microbiome, and the interplay with our immune system. One of the things the "gut" conclusions about hygiene and its supposed dangers does not account for is the coevolution of our immune system and all the bugs around us. When the immune systems of our ancestors hundreds of thousands of years ago adapted to their environment, things were very different from today.

Of course, our hunting and gathering ancestors did get sick. It is impossible to get exact numbers but some scientists estimate that up to one in five people died of infections from pathogens.

But diseases were not the same as they are today. For one, animal parasites were a much bigger deal than nowadays. Head and body lice, ticks, and especially worms were prevalent. Infestations by worms are not a thing most people in developed countries have the decency to be worried about today, while in our past they might have been so common and unavoidable that our immune system had to begrudgingly find a mode of coexistence. But we discussed this already in the last chapter so rest assured, we are done with parasites! Our immune system did not just have to deal with worms though, it also had to arrange with some species of viruses like hepatitis A or bacteria like *Helicobacter pylori* that it just could not eradicate and had to live with.

On top of that, most kinds of diseases we associate with being sick today were practically completely absent in hunter-gatherer communities: Infectious diseases like measles, the flu, and even the common cold. Because most of the worst bacterial and viral pathogens that cause infectious diseases and make our lives miserable in modern times are new to our species in evolutionary terms.

In the world our human immune system evolved in hundreds of thousands of years ago, infectious diseases could not become a major problem because, with a few exceptions, when you survive an infectious disease, you usually don't get it again. Either it kills you or it leaves you completely immune to it for the rest of your life. For the vast majority of human history our species lived in small tribes that were spread thin and, for all intents and purposes, pretty isolated from each other. An infectious disease simply could not become a dangerous threat and establish itself in our ancestors effectively. Because if it infected a tribe, it would infect every available person in no time and then die off because there would be no one left to jump over. So our evolution did not really have to consider these sorts of pathogens that much.

As we became farmers and city dwellers our lifestyle changed forever—and so did the diseases that targeted us. Living close together created a perfect breeding ground for infectious diseases. Suddenly, in evolutionary terms, there were hundreds or even thousands of victims to infect. As our ancestors were not aware of the nature of microorganisms or even basic hygiene and they did not yet possess tools like soap and indoor plumbing, there was not much they could do—on the contrary, their lack of understanding made things worse.

And when they began domesticating and living together with animals in close quarters, often even sleeping in the same rooms, some pathogens jumped over. Our new lifestyle turned out to be the perfect environment for the pathogens of our new animal friends, to adapt to humans and vice versa. As a consequence, virtually every infectious disease we know today arose in the last ten thousand years. From cholera, smallpox, measles, influenza, and the common cold to chicken pox.

And here again, we meet hygiene. Hygiene is incredibly important to protect us from all of these diseases. In the last two hundred years, as we discovered the tiny world with its trillions of inhabitants, we began washing our hands, and we began to clean up and separate our water supply from the places we pooped. We wrapped our food in sterile material and put it in cold places so pathogens could not use it as a shortcut right into our intestines. We started to disinfect things that we use to cut people open and to properly clean up the stuff we use to prepare our food. Hygiene is often confused with cleanliness—but you really should understand it as a targeted approach to remove potentially dangerous microorganisms from the key places and situations where they can make you sick.

Hygiene is a great idea that really benefits the health of our species. This whole point is so important that it is worth repeating: *The microorganisms that are causing infectious diseases are comparatively new to our biology.* Our bodies and immune systems did not have hundreds of thousands of years to evolve alongside them. Surviving the measles does not make you tougher, it just makes your life bad for two weeks. And if your immune system is not in good shape it might also just kill you. *Dangerous pathogens are, well, dangerous.*

Clean water has literally saved hundreds of millions of lives. Hygiene, from washing your hands regularly to making sure your food is properly stored, is incredibly important—as important as vaccines, if not more. Hygiene is also a critical line of defense that keeps us safe from contracting dangerous infections, for example in the case of global pandemics. Coughing into your elbows, washing your hands regularly and properly, and wearing masks buys us time for larger-scale interventions, like vaccines or medication. Hygiene reduces our need to prescribe antibiotics, which automatically combats antibiotic resistance. It protects the weaker members of society like small children and seniors, the immunocompromised, people going through chemotherapy or suffering from genetic defects.

Still, words are important and hygiene and cleanliness are not the same thing. For example, the idea that we scrubbed all microorganisms from our houses and that we live in a sterile world. Nothing could be further from the truth. After you finish scrubbing your floor and carefully wiping down your kitchen and bathroom, your home is teeming with microbes again in no time—even if you used antimicrobial cleaning products. Microbes rule this planet and they rule your house too.

OK, OK, alright. So hygiene is good. But if it is not hygiene that we have to blame, what is the cause of the sharp rise of immune defects in the last fifty years? Well, here it may become counterintuitive because it all has to do with microbes but in a different way. It seems that to train your immune system you need to hang out with *harmless friends.* Your immune system needs nice playdates to learn when to be gentle and forgiving. This more nuanced version of looking at interactions with the microbes around us has been called a few different names, but the nicest one may be the *"Old Friends" Hypothesis,* which focuses much more on our evolution.

For millions of years our bodies and immune systems evolved alongside and together with the organisms that live in the mud and dirt and plants around us. Very early in this book we mentioned that you are a biosphere, surrounded by invaders that want to get in. But you are so much more. You are also an ecosystem where microorganisms of all kinds live together with you. Your body would like to get rid of some but can't and has to learn to coexist with them, others are neutral, and a huge group is directly beneficial to your health. These communities of *commensal microorganisms* are as essential to your survival and health as any of your organs. And one of their most important jobs is to train your immunity.

When you are born, your immune system is like a computer. It has hardware and software and is in theory able to do a lot of things. But it doesn't have a lot of *data*. It needs to learn which programs it needs to run and when. Who is a foe and who can be tolerated. So for the first few years of your life it collects information from its environment. It collects data from the microorganisms it encounters.

It does so by processing "data" it collects from interactions with microbes. If it does not get enough microbial data and can't learn enough, the risk rises that it will grow up to be overly aggressive and will later go on to attack harmless substances like peanuts or pollen from plants.

A really famous study shed some light on how your environment in early childhood forms your immune system. The study looked at two distinct groups of farmers in the United States, the Amish in Indiana and the Hutterites in South Dakota. Both of these populations stemmed from religious minorities that emigrated from Central Europe to the United States in the 1700s and 1800s. Both groups have since not mixed with other populations but stayed genetically isolated, living lives shaped by similar and strong religious convictions. What made these two groups so interesting to study and compare is that both are genetically close, which made it easy to ignore genetics and focus on their lifestyle differences.

And there is a huge difference between Amish and Hutterites: While the Amish practice a traditional style of farming, where single families have their own farms with dairy cows and horses that are used for field work and for transportation and in general avoid modern technology, the Hutterites live on large and industrialized communal farms, with industrial machines and vacuums and many amenities of the modern world. Consequently, researchers found a much higher rate of microbes and microbe poop in the houses of the Amish compared to the Hutterites. The rates of asthma and other allergic disorders are four times higher in Hutterites than in the Amish. So it seems growing up in a less-urban environment offers some protection against allergic disorders.

Also, it is fair to conclude that a little bit of dirt does not harm you, in fact it might be good for you.

Unfortunately (or fortunately, you decide for yourself) most people do not live on farms anymore. Today we don't surround ourselves with the kind of diverse microbial ecosystem that we evolved in parallel to. We isolate ourselves from all kinds of natural environments. There is not one single factor but a bunch of different ones that all play together:

The urbanization of the world has sped up drastically in the last century and in many developed countries the majority of the population lives in cities. And while not all cities are jungles made purely from concrete, the distance to something that resembles nature, with all its critters, makes a big microbial difference. These changes are pretty new in evolutionary terms because until the early 1800s the vast majority of the human population lived in rural areas. This development also coincided with the fact that step-by-step, in the last few decades, through the advent of entertainment and information technologies from TV to the internet, we got used to spending the vast majority of our time inside.

In developed countries, "inside" means an artificial environment made from processed materials that, while not actually sterile, house a completely different ecosystem for a different set of microorganisms than the ones our ancestors adjusted to.

As we said, until very recently in human history, people lived in houses made from natural materials like wood, mud, and thatch, all full of microbes that were all too familiar to our immune systems.

Other important factors are what we put into our bodies. Antibiotic use and overuse was not something our ancestors had to deal with because there were no antibiotics. Not to say that antibiotics are bad—they have created a world for us in which we have forgotten how deadly serious a lot of injuries and infections actually are, because we can just pop a few pills and not die. But antibiotics are not great at discriminating between harmful and helpful bacteria, and so they kill your commensal bacteria, your old friends, too. Aside from the issue of antibiotic resistance in the pathogens we want to kill, overprescription of antibiotics is a huge problem for the healthy microbiome.

The problem may start even earlier, literally at the beginning of life itself: Today a considerable percentage of babies are born via cesarean section. This is not ideal because in regular births the tiny human comes in close and intense contact with the vaginal and often fecal microbiome of their mother. So your birth is actually an important step in the microbial priming of your body and immune system. The microbiome of small children varies significantly depending on how they were born.

Adding another puzzle piece in early life is the fact that fewer mothers are breastfeeding than in the past. Mothers' breast skin and milk contain a vast and diverse array of substances that nurture the very young microbiome and a number of diverse bacteria. Evolution made sure that newborns get plenty of face time with the old and proven microbiome. Both C-sections and the lack of breastfeeding are correlated with a higher rate of immune disorders like allergies.

Maybe one of the most important differences to our evolutionary past may be that modern diets contain vastly less fiber than they used to. Fiber is an important power food for a lot of useful and friendly commensal bacteria, and the fact that we just eat less and less of it means that we can't sustain these little bacteria buddies in the numbers that we might need them.

Phew, OK, that was a lot. Unfortunately there is not one single clean and satisfying answer. The immune system is pretty complicated.

All of these changes in human lifestyle did not show their effects overnight. The transition of our microbial microenvironment and stunted microbiomes was probably a gradual one that happened only in the last century or so. As every generation moved a bit farther away from the natural environment their microbiomes became less diverse and their children inherited their microbiome. Over time the average diversity of the microbiome in developed countries seems to have fallen considerably, especially compared to humans still living a more traditional and rural lifestyle.

All of these factors together probably made for the less than ideal situation we have today. But wherever humans grow up with more access to microorganisms that are old friends, our immune system should fare much better, and indeed there are a lot of observations that support this notion.

Even in developed countries, a number of studies found that children who grow up in the countryside and especially on farms, surrounded by animals and with much more exposure to the outside, suffer significantly less from immune disorders. So while it doesn't seem to make a difference if a house is clean or not, it does make a difference if it is surrounded by cows and trees and bushes and if dogs roam freely.

So what can you take away from this chapter? Wash your hands at least every time you use the restroom, clean your apartment but don't try to sterilize it, and clean the tools you use to prepare food properly.

But let your kids play in the forest.

42 How to Boost Your Immune System

By now, the immune system has hopefully lost some of its cloudy and mystic aspects to you. It is not a magic force that can be charged up like an energy shield or laser weapon, but a complex dance of billions of parts. A beautiful symphony that has to follow a strict choreography to function in harmony. Any deviation makes your immune response either too weak or too powerful, and neither is good for your well-being and survival. If you read this far you already know more about immunology than 99% of the general population. So think about it: if you could—which parts of your immune system would you like to boost?

Would you like more aggressive and stronger Macrophages or Neutrophils? Well, this would mean more and stronger inflammation, more fever, more feeling sick and tired even if you encounter only small and minor infections. How about superstrong Natural Killer Cells to kill more infected or cancer cells? OK, but they might be so motivated that they eat away at healthy cells that just happen to be around!

Do you want to boost your Dendritic Cells so that they will start activating the Adaptive Immune System more? That would drain and exhaust the resources of your immune system even for small dangers, leaving you open and vulnerable when a seriously dangerous infection happens.

Or maybe you could boost your T and B Cells, making them much easier to activate, but which would cause autoimmune diseases as some of these cells will certainly begin attacking your own tissue. Once your boosted antibodies and T Cells have begun killing your heart or liver cells they will not stop until they have finished their job.

Maybe this is all not dangerous enough for you, and you would rather boost your Mast Cells and the B Cells that produce IgE Antibodies, the combination of cells that is responsible for allergies. Food that was only mildly irritating to your bowel will now cause violent diarrhea or allergic reactions that could kill you in minutes.

Is all of this too boring? Why not think outside of the box and boost all the regulatory parts of your defense systems so they shut your immune system down, leaving you open for infections even by the most harmless pathogens? You probably understand what I'm getting at here: *Boosting the Immune System is a horrible idea that is used by people trying to make you buy useless stuff!*

Luckily the danger is pretty mild that you could actually boost your immune system since basically nothing you can buy legally actually does it! Even the mere term "strong immune system" is a misnomer. Over everything else, you want a *balanced* immune system. *Homeostasis. Aggression and calmness.* You want elegant dancers who remember the choreography really well over pumped-up rugby players who want to smash stuff. In all likelihood, your immune system works exactly as intended.

OK now, wait. If boosting the immune system is so crazy complicated and dangerous, why is the internet filled with products that promise to do exactly that?

From infused coffee to protein powder, mystical roots dug up in the Amazon rainforest, or vitamin pills, there are an endless amount of things you can buy to "boost" your immune system.

In reality, nobody knows how many cells of which kinds and at what activity level are necessary to have your specific immune system work optimally. Anybody who says that they know what is needed is probably trying to sell you something.

At least for now, there are no scientifically proven ways to directly boost your immune system with any products that are easily available. And if there were, it would be very dangerous to use them without medical supervision.

The most important thing you need to do to have a healthy immune system is to eat a diet that provides you with all the vitamins and nutrients your body needs. The reason is, simply, your immune system constantly makes many billions of new cells. And all these newborn cells need resources to function properly. Malnutrition is strongly associated with a weak immune system. If you are starving, you are more susceptible to infections and diseases because your body has to make hard decisions and the immune system suffers from that.

But if you eat at least somewhat of a balanced diet with some fruits and vegetables, you will get all the micro and macronutrients for your immune system to work just fine. Interestingly, even in developed countries there is micronutrient malnutrition, especially among elderly people. Which just means that someone has a deficiency of essential nutrients and vitamins—usually because they don't eat enough or have too little variety in their diets. So only eating pizza is not healthy but this should already have been clear. In all likelihood if you eat sort of OK, your immune system works just as it should.

On top of just eating right, the positive health effects of even moderate regular exercise have been known for a long time. Your body is made for movement and so moving it around a bit keeps a variety of systems in good health, especially your cardiovascular system. Working out also directly boosts your immune system, because it promotes good circulations of fluids throughout your body. In a nutshell, just by moving, stretching, and squashing your various body parts, your fluids flow better and more freely than if you lie on your couch all day. And good circulation is good for the immune system because it allows your cells and immune proteins to move more efficiently and freely, which makes them do their job better.

But that is what you can do basically.

Some people actually do have deficits and benefit from certain supplements but this is nothing you can self-diagnose. The bitter reality is that humans are very different and the reasons why different changes in diet or lifestyle may have a positive or negative effect on you are way too complex to summarize in a general book about the immune system.

If you feel like you have a shortage of a vitamin or microelement or something, this is something you should discuss with your doctor in real life.

This blanket statement will leave many people unsatisfied. How is it possible that humans can fly to the moon, that we can build particle accelerators, and that we came up with 980 different Pokémon but we can't improve our immune system?

Well, look at it like this: If you have an old rusty car that you have used for off-roading for decades, with a broken axle, blown-out tires, and a busted headlight, do you think you could fix it by putting special gas in its tank and slapping on a new finish? You can't undo the damage that you caused by treating a car really badly through magic. If you want your car to run better and longer, just take care of it and, you may have guessed it, your body is exactly the same.

If you want to "boost" your immune system so that it is healthy, start by taking better care of yourself by living a healthy lifestyle, and the complex concert of your immune system, with all its billions of different parts, will run properly for a longer time. Not forever unfortunately, neither cars nor humans are made for that. But longer and better. This is what science has to say about this topic, at least for now.

Talking about boosting the immune system and the unscientific claims made by many people working in the multibillion-dollar industry of selling supplements would be mildly funny, as the majority of people at worst are wasting money. Unfortunately there are millions of people who suffer from real and serious diseases that are anything but funny, from cancer to autoimmunity.

And these people, who often are desperate to alleviate their symptoms or are just frankly trying to survive, are the ones who might fall victim to the empty promises of the supplement industry. Even worse, some might even go as far as to disregard real medical treatment as a consequence of these lies supported by greed or well-meaning but ill-advised appeals to naturalism. These broken ideas about health and boosting the immune system can be perpetuated only by our collective misunderstanding of the mechanisms of our own immune system and what it actually is.

Even experts have to be very careful if they want to attempt to boost the immune system and this might be the right moment to tell a story where it all went horribly wrong.

As the knowledge of the mechanisms of our immune system grew massively in the last few decades, scientists tried to come up with new ways to fight the diseases that haunt us. If we could just manipulate our intricate defense system the benefits to our species would be enormous. But as we said, manipulating the immune system is very dangerous. It is constantly performing a balancing act between too harsh and too mild and trying to interfere can go horribly wrong.

An infamous example is TGN1412, a drug trial that went so horribly wrong that it reached beyond the sphere of immunology and got some headlines in newspapers. The trial was supposed to look for side effects in humans taking a drug that was supposed to stimulate T Cells in cancer patients and make them survive longer.

The drug was an artificial antibody that was able to connect to and stimulate the CD28 molecule on T Cells—we already met CD28 before without naming it—one of the signals T Cells need to be activated. We described it as a gentle kiss the Dendritic Cell needs to give a T Cell to activate it.

So the idea of TGN1412 is pretty straightforward: Give T Cells an artificial "kiss" to stimulate them to be more effective and easier to activate in cancer patients. Basically "boosting" their immune system to be more formidable in the face of this life-threatening disease. And yeah, well, boost it did.

For safety reasons, the amount of TGN1412 administered was 500 times lower than the dose that had led to any reaction in macaques (which is a cute species of monkeys, if you are wondering) and so the researchers doing the drug trials did not expect any real reaction to occur in the human volunteers.

But instead, minutes after TGN1412 was given to healthy young men, all hell broke loose. It turned out macaques, the animals used to test the drug in animal models, happen to have way fewer CD28 molecules on their T Cells than humans and so they reacted way less strongly to the drug than expected, creating a false sense of security. Also, for some reason the drug was administered ten times faster to the human volunteers than it had been in the animal model.*

---

* Somewhere in this book this needs to be mentioned and we might as well do it here. Beware headlines about anything health related that mentions animal models. Yes, it is crucially important to test drugs in animals, but unsurprisingly, animals and humans are different. Yes, we have created mice with immune systems that basically mirror our human immune system, yes, we have monkeys like macaques that live on an evolutionary branch that is not that far away from ours, but these are still completely different organisms. There are all kinds of drugs that heal mice or prolong their lives and do whatnot—but do nothing at all for humans. Or worse, they are dangerous or even deadly for us. So again, this is not saying that these experiments are not critically important, unbelievably important knowledge has been gained through animal models. But when

Within minutes the volunteers experienced an extremely strong cytokine release syndrome, which is a cytokine storm on speed. All over their bodies, billions of immune cells that usually require careful activation, guarded by the safeguards we discussed in this book, all awoke at once. Basically, all the T Cells in the volunteers' bodies were overstimulated and released an onslaught of activating and inflammatory cytokines. These great floods of cytokines activated more immune cells, which then themselves released more cytokines and caused more inflammation. A horrible chain reaction that was self-perpetuating and accelerating.

The immune system of the volunteers had been released and nobody was prepared for what was happening. In a rapid, violent, and systemic reaction, fluid rushed from their blood into tissue all over their bodies, making them swell up as they were writhing in excruciating pain. What followed was multiorgan failure and the volunteers could only be kept alive by machines and huge doses of drugs that shut down the immune system. One of the worst-affected volunteers suffered from heart, liver, and kidney failure all at once and would later go on to lose many toes and some fingertips. Luckily, all of the six volunteers survived this horrible day, and after a few weeks of intensive care, most of them were able to leave the hospital again.

As the TGN1412 trial failed in the most horrible way possible, it obviously sent shockwaves through the medical research community. Many guidelines for human trials were amended as a consequence of this.

OK, so what is the purpose of this horror story? Certainly not to say that drugs that boost the immune system are a bad idea in general, but it teaches us about the complexity and dangers of using them. If we look at the scale and mind-bending level of detail and complex interactions of your immune system, the magnitude of the challenge to manipulate it becomes clear. Make no mistake—while we did discuss a lot of things in this book, I still massively simplified everything. We barely scratched the surface from the perspective of actual immunologists that are working in the trenches of immunology.

---

it comes to drugs and cures, everything can be different once a drug is used by humans. So if you hear news about some kind of amazing drug, make sure to check if the excitement is based on human trials or if it is still in an earlier stage, tested in animals.

Think of the immune system as a crazy large machine with thousands of levers and hundreds of dials. Billions of gears and screws and wheels and blinking lights that are interacting inside it constantly. Pull any lever and you just don't know for sure what sort of downstream interactions you will cause.

OK, so boosting and strengthening the immune system is complicated for experts, and outside of a healthy lifestyle is pretty impossible (and ill-advised) for regular people. But there is actually a huge thing you could do to at least prevent damage. It turns out that many people are actually suppressing their immune system without being aware of it.

43 Stress and the Immune System

To understand the role of *stress* and the immune system, we have to look back millions of years, to a simpler but much more cruel time in our developmental history. In order to survive, your ancestors had to deal with the evolutionary pressures the environment subjected them to. In the wild, stress is usually connected with existential danger, like a rival that crosses into your territory or a predator that wants to make you its meal.

So for your ancestors it was a good idea to react strongly to perceived danger because if you acted decisively, then you were more likely to survive—if you were wrong and something was not actually dangerous nothing was lost. If you were slow to react to potential danger and you were wrong and it turned out to be actually dangerous, something bigger would probably eat you. As a consequence, organisms that were good at responding rapidly to a possible source of danger, a *stressor*, real or not, were more successful at surviving and reproducing than others who were not.

Over time and through this selective pressure, our ancestors became fine-tuned to recognize stressors quickly and to react swiftly to them, often with automated processes. In mammals for example, this meant glands that are able to rapidly release stress hormones, which accelerated the delivery of oxygen and sugar to the heart and skeletal muscles and made it possible to react to a threat instantly and with power. Behavioral adaptations like the fight-or-flight response saved even more crucial time and helped them to survive in the wild. Because if you think you have spotted a lion in your peripheral vision, it is a better survival strategy to start running or to throw your spear than to carefully consider for a minute if that really was a lion or just a bush that sorta looked like one.

In the context of these sorts of adaptations, it makes sense that your immune system also responds to stress. No matter if you fight or flee, in both cases, the likelihood that you will get wounded increases dramatically, which means that pathogenic microorganisms might have an opportunity to infect you, making your immune system immediately relevant. So one of the adaptations to stress was to accelerate certain immune mechanisms while slowing down others.

Now we can call ourselves incredibly lucky that we left the lifestyle of our ancestors behind us and invented civilization and food delivery and comfy homes and that we killed all the large things that tried to eat us (the small things that are still trying to eat us are a bit harder to manage, unfortunately). But despite all of these great inventions, our bodies unfortunately did not get the memo yet. They still behave as if we were trying to survive in the savannah or as if we had to deal with wild lions hunting us on a regular basis. And so they still hold on to as many calories as possible, despite the abundance of food in the modern world. And they are triggering a stress response in situations that actually require calmness and clear thinking. Running away will not help you to pass the exam tomorrow. You can't physically fight your client if the deadline is close (OK, technically you can, but it probably won't help you). Our body does not know that though, and thus this unfortunate misunderstanding causes stress. Psychological stress has actual and immediate physical consequences for the immune system, many of them not helpful.

The thing about stress is that it is similar to your immune response in one extremely important aspect: When it works as it is intended to, stress is a great mechanism that helps solve an immediate problem and then shuts itself off afterwards. But the nature of the stressors we encounter in the modern world is different than the ones we evolved with. In the past the lion either got you or you escaped—either way, your stress stopped. It rarely followed you around for weeks or months, like exam season or a large project for a demanding client. And so a mechanism that was meant to support short bursts of activity has turned into a chronic background noise.

So what is the effect of chronic stress on your immune system? Well, as so often before, it's very complicated and not at all straightforward. When we talk about stress and its effect on health we open up topics like depression, loneliness, specific life situations, and the different ways people deal with them. As soon as your behavior is involved, things become hard and fuzzy. You can't just say that chronic stress causes autoimmune diseases because it could be and almost certainly is more nuanced.

For example, we know that stress can be one of the factors that lead people to smoke more cigarettes. And smoking is a risk factor for autoimmune diseases like arthritis. So we need to be very careful with our words in this section because a lot of uncertainties exist here! Having gotten this disclaimer out of the way, it is clear that chronic stress is very unhealthy and related to a number of diseases and conditions.

In general, chronic stress seems to disrupt the ability of the body to shut down inflammation and causes chronic inflammation. And as we discussed before, chronic inflammation has been linked to a higher risk for numerous diseases, from cancer to diabetes, heart and autoimmune diseases, and also a general frailty and higher chance of death. Chronic stress changes the behavior of your Helper T, which is not great since they are particularly important conductors and influence a number of other immune responses. This can lead your Helper T into making the wrong decisions, which can throw your immune response out of balance.

Stress also releases hormones like cortisol that shut down and suppress your immune system, making it weaker and less able to do its job properly in a variety of ways. Wounds heal slower, infections are more likely to break out and to cause disease. Already present pathogens or diseases can no longer be held in check efficiently, leading to an outbreak of herpes, for example. Or in more serious cases, to a much faster progression of HIV. Chronic stress means a chronic release of cortisol, which generally slows your defense systems down.*

A pretty strong connection has also been made in recent years between the onset of autoimmune diseases and stress. And stress also seems to be one of many risk factors for tumor progression.

OK, so this range of possible diseases could not possibly be broader—it

---

* This is an issue in professions that are very taxing for the body, like elite Special Forces units in the military or professional athletes. One downside of these kinds of jobs is higher levels of cortisol and lower levels of antibodies and important cytokines.

seems that chronic stress negatively affects every single area where the immune system is supposed to protect you.

So if you are still looking for ways to boost your immune system, an actual tangible thing you can start doing today is to try to eliminate stressors in your life and to take care of your mental health. This might seem like pretty daft advice because it is so obvious, but the connection between your state of mind and your health is very real. So helping people live happy and fulfilled lives with less stress and depression would probably have considerable health benefits for our societies.

44 Cancer and the Immune System

To many people cancer is probably the scariest disease that exists. Even just saying its name triggers terrified feelings in some. It is the greatest betrayal you can experience: your own cells deciding that they no longer want to be part of you.

In a nutshell, cancer is when cells in a certain part of your body begin to grow and multiply uncontrollably. There are basically two major categories: When cancer cells form in solid tissue, like your lungs, muscles, brain, bones, or sexual organs, they form *tumors*. You can imagine tumors basically like cells starting a new small village that eventually grows into a metropolitan area sprawling across the continent that is your body.

The word tumor originally just means "swelling," and just like a swollen body part, a tumor is not automatically a deadly disease. There are so-called "benign tumors," which are like the confused cousins of cancer. The main difference is that benign tumors do not invade other organ systems, like cancer cells. They basically stay with their friends and just grow as a physical mass inside your body. And so the outcome with these tumors are very good—often they only need to be monitored rather than destroyed or treated. Even benign tumors can become dangerous though, if they grow too large and begin to press on organs like your brain or affect vital systems like your blood vessels and nerves. In these cases usually the tumors are removed with as little damage to the surrounding tissue as possible. So yeah, tumors suck in any case, but if you have to pick one, benign tumors are the way to go.

In contrast to tumor-forming solid cancers, "liquid" cancers affect your blood, bone marrow, lymph, and lymphatic system and often start in your bone marrow, and what basically happens here is that the superhighways of your vascular and lymphatic systems are overwhelmed and crowded out by useless cancer cells. (Liquid cancers still are made from cells, they are not actually liquid.) Leukemia, or blood cancer, is often used as a sort of catchall name for these kinds of cancers.

Cancer can emerge from basically every type of tissue and cell in your body. And since you are made up of many different types of cells, it is not really a single type of cancer, but hundreds of different ones. Each of them is special and with its own individual challenges. Some are very slow and can be treated well, while others are very aggressive and extremely deadly. Almost one in four people alive today will get cancer during their lifetime. And one in six people will be killed by it. So basically everybody will get to know someone who has to deal with the disease at some point in their life.

Despite the horrible harm they cause, cancer cells are not evil. They don't want to hurt you. They don't want anything, really. As we established, cells are protein robots that just follow their programming, which unfortunately can be broken and corrupted.

Or not them, but their programming. To make a long story short, your DNA carries the code of life, building instructions for all the proteins and parts that make up your cells. These building instructions are copied and transferred from your DNA to your protein production machines, your ribosomes, where they are turned into proteins. And the number and production cycle of different proteins enable your cell to do different things, like sustain itself, react to stimulation, or behave in certain ways.

Because this process is so central to life, if your genetic code gets damaged, it will have consequences down the line. Maybe some proteins will not be correctly built, or too many or too little of them, all of which affects how well your cells work. These changes in your DNA are called mutations—and while this is quite a loaded word, it basically just means that your code changed a bit. Now, your DNA is damaged and changed all the time, every second of your life. The genetic code in an average cell is damaged tens of thousands of times a day, which means in total you suffer from trillions of tiny mutations each day. This sounds worse than it is, as almost all of them are either fixed very quickly or not problematic. So most of the mutations accumulating will be of little consequence to you.

Still, over time this means that damages are accumulating, just as a consequence of being alive and having your cells multiply. Remember in school when some teachers handed out terrible photocopies of worksheets that were already a bit fuzzy around the edges? Imagine having to make copies from copies of copies. Over and over again, over years, maybe decades. Maybe one day a hair got on the scanner or a corner got frayed. These mistakes became part of the new copies and therefore of all the copies that followed after that.

In your cells most of this damage just happens through the basic process of living life, through your cells dividing and keeping the body going, without any special reason or cause. It is just statistics and bad luck. You can do a lot to increase your chances of getting cancer through your lifestyle by doing things that damage your genetic code, like smoking cigarettes, drinking alcohol, or being obese, through contact with carcinogenic substances like asbestos, or simply by enjoying beautiful summer days without sunscreen.*

---

* There is this myth that your attitude is crucial when it comes to surviving cancer. The general idea is that if you have and display a positive attitude, you will activate some mystical force in the immune system and enable it to overcome the disease. Inversely, a really negative attitude may have the opposite effect and make it harder for your body to beat the disease or may even have caused it. Wherever the idea of your attitude affecting your cancer survival chances originally comes from, after decades of research it has become clear that with an extremely high certainty, your attitude has no effect on your chances of surviving cancer. Your immune system does not magically become better or worse at fighting cancer if you are or are not positive and happy. Still, this myth is going strong, as it appeals to our culture of self-empowerment and agency and is spread by many well-meaning people.

But aside from the fact that there is no solid science to prove a relationship, it is a terrible thing to say to someone with cancer that their attitude matters and that they should stay positive, because it does two things:

For one, it puts the responsibility of healing and surviving on the sick person. It implies that if you don't win the fight and have faced the gravest of all outcomes, it is your fault. That if you just had been more positive and optimistic, no matter how you really felt, that you could have saved yourself. Which is an incredibly unfair burden to put on someone who is fighting this disease.

The other reason is chemotherapy, surgeries, and radiation therapy are, well, not a great experience. And by being told that you are supposed to be positive to get well, you are told that you are not allowed to feel how you feel. But expressing how unwell you are and asking for an open ear or love is important because it can help you deal with very strong negative emotions caused by fear and very unpleasant treatments that you have to endure. Being more positive and having a good attitude towards life and its challenges makes your life better. It does so no matter if you are sick or not—if you have more good and optimistic feelings, you feel better. It can reduce stress, which in turn might reduce the negative influence to your immune defenses. So a positive attitude when you are sick is a good thing. Studies have shown that a positive attitude during

All in all, the easiest way to get cancer is to be alive long enough. It is statistically impossible to not develop some cancer at some point in your life, even if it ends up not being the cause of your death.

To become cancer, a cell has to mutate in just the right way to acquire specific corruptions in three different important systems that work in tandem to prevent cancer.

The first key mutation has to appear in *oncogenes*, genes monitoring the growth and proliferation of the cell. For example, some of these genes were very active when you were an embryo, a small clump of cells. To turn a single original cell into trillions in mere months, it needs to divide and grow rapidly, to eventually become a tiny body. These rapid growth genes are turned off eventually, when there is enough of you to form a somewhat complete human. Years or decades later, when a mutation switches these oncogenes on again, the corrupted cell can begin to divide and proliferate rapidly, just like when it was trying to create a new human inside a womb. So mutation number one: Rapid growth.

The second key mutation has to happen in the genes that are responsible for fixing your broken genetic code, appropriately named *tumor suppressor genes*. These genes produce safeguards and control mechanisms that continuously scan your DNA for mistakes and copying errors and fix them right away. So if these genes are corrupted or faulty, your cells basically lose the ability to repair themselves.

But these two specific mutations are still not enough.

Cells usually recognize when their code becomes dangerously broken and they are at risk of going rogue. If they notice in time, they trigger their own destruction and kill themselves. So the last group of genes that needs to be corrupted are the genes that make a cell commit controlled suicide by apoptosis. We talked about apoptosis a few times already: it is the way most of your cells end their own lives—a constant process of self-recycling that prevents your cells from amassing too many mistakes over time.

If cells lose the ability to kill themselves when it is time, when they become unable to fix the mistakes that are naturally amassing in their genetic

cancer treatment is good for your own mental well-being. It can make the experience much less worse. And less worse is a very good thing during chemotherapy.

code, and when they begin to grow without restraint, they become cancerous and dangerous. Of course, we did simplify a little bit here. A single mutation in these three systems is usually not enough. Multiple genes in each of these systems have to mutate in a bad way. But this is the basic principle underlying cancer.

In a sense, once these damages amass and one of your cells turns into a cancer cell, it becomes something else. Something ancient and something new. Over billions of years, evolution molded cells to optimize themselves to survive and thrive in a hostile environment. Fighting each other for resources and space. Until a very new and exciting way of life emerged: Cooperation. A form of cooperation that allowed for a division of labor and allowed cells to specialize and to become more successful as a group. But cooperation required sacrifices. For a multicellular being to be able to stay alive, the cohesion and the well-being of the collective has to matter more than the survival of the individual cell.

Cancer cells walk this process back and stop being part of the collective and in some sense, become individuals again. And that would be OK in principle. Your body can handle a few cells doing their own thing and even live in harmony with them. But unfortunately cancer cells usually do not content themselves with doing their own thing, but divide and divide, again, and again. They turn from being an individual to a collective again. A sort of new organism within you. Still part of you but also not you at all. They take the resources you need to survive, destroying the organ systems they used to be a part of, and begin competing for the space that you inhabit.

One might think that evolution should have taken care of this sort of corruption, but since you tend to get cancer after leaving the reproductive age, there was little incentive to optimize for good protection against cancer. In 2017 only 12% of all cancer deaths happened to people younger than fifty. And so, if you are lucky enough to get old, you are almost certain to have some amount of cancer cells inside you, it might just be that other things kill you before those get a chance.

Because cancer is a constant danger and an existential threat to survival, the human body in general is actually pretty good at dealing with it. Or more precisely, your immune system is. It is almost certain that your immune cells killed a bunch of cancer cells somewhere in your body while you read the last few chapters.

Over your whole life, some of your cancer cells may have even grown into small tumors that were eventually wiped out by your defenses. This might have happened today and you would have no idea that it happened. So you can rest assured that the vast majority of cancer cells you develop in your life will be killed without you even noticing. And while this is great, we don't care about the 99.99% of occasions where things went fine, we care about the one time where the immune system is overcome and a young cancer cell becomes a proper, life-threatening tumor.

So let us take a look at *immunoediting*, the back-and-forth, the struggle between your immune system and the cancer cells. In general it goes like this:

1. The Elimination Phase

So congratulations, you have a proper cancer cell. It is no longer able to monitor and repair its genetic code, it can't kill itself anymore, and it has lost its restraint and is beginning to multiply rapidly. And it mutates more with each generation. Not great, not terrible.

Over a few weeks the cell clones itself uncontrollably, creating at first thousands, then tens of thousands of copies into a tiny, tiny patch of cancer. This rapid growth needs a lot of nutrients and resources. So the mini tumor begins stealing the nutrients from your body, by ordering the growth of new blood vessels made only to nourish it. And so the cancer cells cause damage by behaving egotistically. In its neighborhood, healthy body cells begin to starve and die.

But as we learned before, the unnatural death of civilians attracts attention, as it causes inflammation and puts your immune system into ultrahigh alert.

Let us paint a picture that illustrates what happens here: Imagine a group of people in Brooklyn decided that they were no longer part of New York City but that they were now a new settlement called *Tumor Town* (subtle, I know) that just happens to occupy the same space.

The new city council of Tumor Town is ambitious and wants to create an amazing new town center, so it orders tons of construction materials like steel beams, cement, slabs, and drywall and just begins building new apartment buildings, convenience stores, and industry right in the middle of the place formerly known as Brooklyn. None of the new buildings and structures built to code, of course—they are badly planned, brittle, and dangerous, with sharp edges and dangerously crooked. Also they look pretty ugly. There is also no apparent logic to all of it, the new buildings are built right in the middle of streets and on top of playgrounds and on existing infrastructure. To connect all of the new construction the old neighborhood is torn down or overbuilt to make room for new highways and divert traffic and tourists from New York to Tumor Town. Many of the former residents of Brooklyn are trapped right in the middle of it. Some grannies are firmly walled in, have no way to get groceries, and begin to starve.

This goes on for a while until one day, alerted by a lot of complaints about the stench of dead grannies, NYC building inspectors and police show up, looking for the people doing the construction.

If we bring this back to your body, attracted by the commotion caused by the uncontrollably growing cancer, the first immune cells will make their way to the tumor and invade it: Macrophages and Natural Killers want to see what is going on. Now, a hallmark of cancer cells is that they show signs of being "unwell." Like they don't have display windows or just have a lot of stress molecules on their membranes, so Natural Killer Cells go right to work, killing cancer cells and releasing cytokines that cause more inflammation, while Macrophages clean up the bodies.

Through the signals from the Natural Killer Cells, Dendritic Cells realize that something dangerous is present and activate into danger mode. They collect samples of dead cancer cells and begin activating Helper and Killer T Cells in the lymph nodes. You are probably wondering how the adaptive immune system can have weapons against cancer cells since they are part of the body.

As we said in the beginning, cancer cells always come with a certain set of genetic corruptions, which lead to corrupted proteins. Some of your adaptive immune cells have receptors that are able to connect to these proteins. In any case, by the time your adaptive immune cells arrive, the tumor has grown to hundreds of thousands of cells but this is about to change. T Cells begin by blocking the growth of new blood vessels, which starves many of the cancer cells outright or at least makes it very hard for the tumor to grow further. Imagine the building inspectors in Tumor Town putting up roadblocks and ending the transfer of tourists and resources into the new illegal city.

The Killer T Cells scan the tumor cells' display windows for malformed proteins that should not be there and order them to kill themselves. Natural Killer Cells kill the cancer cells that have hidden their MHC molecule windows. With no possibility of hiding and no way to order fresh nutrients from the blood, the tumor collapses. It is a massacre and hundreds of thousands of cancer cells perish. Their carcasses are cleaned up and consumed by Macrophages. Imagine that as New York City would tear down illegal construction, your body crushes the illegal tumor. Except . . . something did not go as planned.

2. Equilibrium

While the battle seemed to be over, natural selection spoils your sweet victory. The initial response of the immune system was very effective. Your immune cells killed the cancer cells that were so kind to let them know that something was seriously wrong with them. Which is exactly how your cells are set up—they are supposed to signal that they are broken—it is actually a sign that they are not completely corrupted yet. Under normal circumstances this is enough and the tumor is eliminated.

But if things go wrong, the cancer cells have time to corrupt even more, a little bit like the viruses that we met before. As they multiply rapidly and unchecked, there are more opportunities for new mistakes in their genetic code to appear, especially since their self-repair mechanisms are already damaged.

The longer these cancer cells are alive and the more they proliferate, the

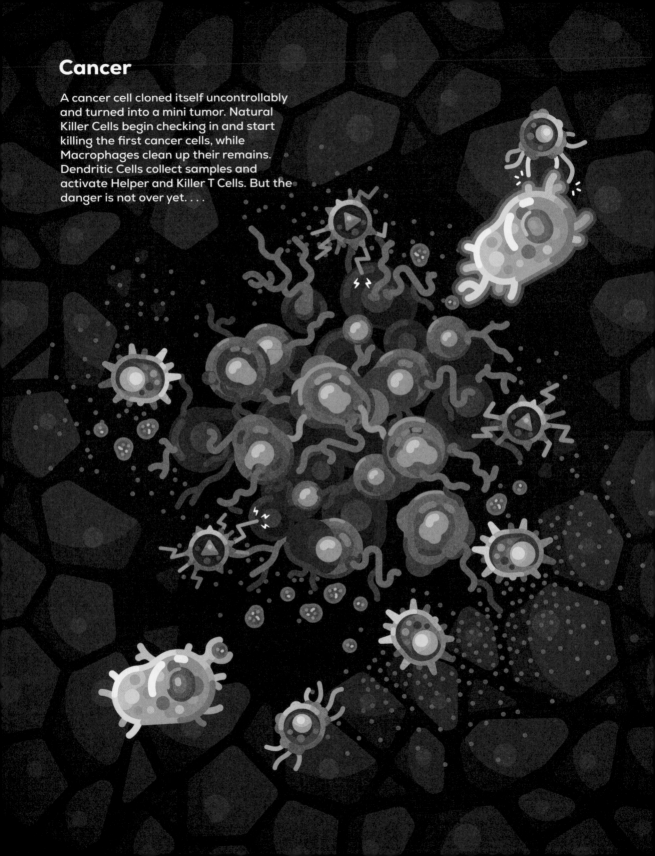

Cancer

A cancer cell cloned itself uncontrollably
and turned into a mini tumor. Natural
Killer Cells begin checking in and start
killing the first cancer cells, while
Macrophages clean up their remains.
Dendritic Cells collect samples and
activate Helper and Killer T Cells. But the
danger is not over yet. . . .

higher the chances that they acquire new mutations that turn out to make them a tiny bit better at hiding from the immune system. And as evolution goes, by doing its best to destroy the cancer, your immune system is selecting the most fit cancer cells. In the end, hundreds of thousands of cancer cells died. Maybe even millions. But a single cancer cell is still here and has found ways to fight back efficiently.

For example, one of the genius and horrifying methods cancer cells have to protect themselves from the immune system is targeting *inhibitor receptors* on Killer T Cells and on Natural Killer Cells. Inhibitor receptors inhibit these cells from, well, killing. They are a sort of off switch that deactivates Killer Cells before they can attack a cell and destroy it—which in principle is a good idea. We've hit on the fact of how dangerous your immune system is many times, and there need to be mechanisms to stop overeager immune cells, so inhibitor receptors play an important role in the complex concert of your immune system. But unfortunately cancer cells can mutate in a way where they are able to shut down your Killer Cells.

So now we have a cancer cell that is able to shut down the defenses of the immune system. And so a new tumor begins growing, producing thousands of new clones that are changing and mutating again.

3. Escape

The new cancer cells that were molded and formed by the countermeasures of your immune system are the ones that will eventually cause all the trouble. In a perverse way they become immune to the immune system. They do not show their broken nature on their surfaces. They do not release too many signals that alarm the body. They are quiet while hiding in plain sight. They are actively shutting down the immune system by sending corrupt signals. And they are growing. As the tumor is expanding it again begins killing healthy tissue and this is attracting attention—but this time the tumor is no longer a pushover. The final phase begins: Escape.

The cancer cells begin creating their own world, the *cancer microenvironment.*

If we think back to Tumor Town in Brooklyn, everything is different this time. The town has been rebuilt but now the new city council has forged all sorts of permits that confuse the NYC building inspectors. No longer are they able to order the destruction of the sprawling Tumor Town that is slowly taking over the city. This time new roadblocks make sure no inspectors can enter the rapidly growing, illegal settlement and check if the fake permits are correct. The cancer cells have created a sort of borderland that is hard to cross for your immune cells.

If all of these things come together the cancer has basically won and tamed the immune system successfully. All avenues of attack have been shut down and uncontrollable growth is the consequence. In the end, if left untreated, these new and optimized cancer cells become metastatic, which means that they want to explore the world and expand into other tissues or organs, where they continue growing. If this affects vital organs like the lungs, the brain, or the liver, the intricate and complex machine that is your body begins breaking down.

Imagine installing new but useless parts into your car engine every day— your car will work for a while but at some point the engine will no longer start. This is how cancer kills you in the end. By taking up so much space and stealing so many nutrients that your true self has no room to properly function anymore and your affected organs have to shut down. This, in a nutshell, is how cancer overcomes your immune system. Although, as we'll discuss in the last chapter of this book, your immune system might also be the key to successfully overcoming cancer, or at least to make it much less deadly.

But for now, since we are already talking about cancer, let us look at something you can do to actively increase your chances of getting it and the role your immune system plays here!

An Aside Smoking and the Immune System

WHILE AIR POLLUTION IS A THING AND RESPONSIBLE FOR UP TO FIVE MIL-
lion deaths per year, none of the things you can breathe in by strolling

through a city compares even mildly to what you get by smoking a single cigarette. While you might know that smoking is incredibly bad for you because "something-something cancer," there is more! It turns out smoking is bad in many different ways that are closely related to the immune system. In a nutshell, you break the mechanisms that protect you against disease and cancer, while making it more likely to get an infection or cancerous cells!

Cigarette smoke is saturated with over 4,000 different chemicals, many of them with unknown properties and interactions with each other. But we know for sure that nicotine, the magic and vile substance that makes smoking addictive, suppresses your immune system. It makes your immune cells slow and ineffective. The major place where this happens is the respiratory system, especially in the lungs—which should be no surprise because this is where all the smoke goes. What does nicotine do exactly?

First of all, it affects your Alveolar Macrophages we met briefly before. They are basically just Macrophages that are more chill and patrol the surface of your lungs to pick up trash and the occasional pathogen. In the lungs of smokers there are way, way more of these special Macrophages than in nonsmoking people. Which makes sense, because cigarette smoke comes with all sorts of microparticles and charming stuff like tar that need to be cleaned up constantly. But because of the constant exposure to nicotine, these already toned-down Macrophages are toned down even more. No longer just chill but basically constantly tired and sluggish.

They are less able to call for support and reinforcements and they have a much harder time killing enemies. On top of that, these poor dysfunctional Macrophages are also accidentally damaging your lungs by periodically vomiting out chemicals that dissolve your lung tissue.

Given enough time these Macrophages high on nicotine can destroy large amounts of the functional tissue of the lung, creating wounds that turn into scar tissue. If this is not clear from the context: Scar tissue in your lungs is bad if you like breathing. Wounds in the lungs also come with the unfortunate side effect of inflammation, which activates more immune cells, which then causes further damage.

Another crucial cell that becomes seriously toned down and less active through smoking are your Natural Killer Cells, which as we learned before are one of your main countermeasures against young cancer cells. It is thought that this plays a relevant role in the considerably higher incidence of lung cancer in smokers. Which makes sense—on the one hand you saturate your lungs with cancerous poison and a drug that makes your immune system cause wounds in your lungs, and on the other hand you make the cells tasked to kill cancer less able to do so effectively.

What about the adaptive immune system? Although regular smokers have many more immune cells in their blood overall, they seem to be less effective. T Cells have a much harder time proliferating after they have been activated and their behavior is also more sluggish. Antibodies in general seem to decay much faster in the bodily fluids of smokers so the overall effectiveness of the adaptive immune system is greatly reduced, which explains why infections like the flu are much more deadly for smokers.

Although there is an exception: Autoantibodies, a kind of antibody that can cause certain autoimmune diseases, are greatly increased. In a nutshell, if you smoke, your immune system does way more of the stuff that is bad and damaging for the body and at the same time it is worse at actually fighting your enemies, calling reinforcements, and stopping invaders from spreading. As a bonus consequence smokers have a harder time healing wounds because of their suppressed immune system not being able to help with healing as much as it should. Even if you stop smoking today, your immune system will stay suppressed for a week to months—so the earlier you quit the better.

But it would be dishonest to say that there are not a few positive effects of smoking because the world is not binary: Sometimes having a toned-down immune system can be a good thing. Inflammation is a double-edged sword, indispensable for your survival, but also very harmful to you.

Smokers suffer less often from inflammatory diseases simply because inflammation is regulated down if your immune system is behaving like a slug that ate a few hash brownies for breakfast. So in the case of certain inflammatory autoimmune diseases, like ulcerative colitis for example, smoking seems to offer some form of protection.

Don't use this fact as an argument in your next discussion with your mom why you should totally continue to smoke though: All in all, while

smoking offers some protection against some diseases, it also makes you much more susceptible to many, many more others. The slight upsides are not worth the massive downsides. This would be a great place for an analogy about how stupid it would be to smoke to avoid certain diseases but actually, maybe this is already the analogy. Smoking to have a slightly better chance to avoid inflammatory diseases would be really stupid.

45 The Coronavirus Pandemic

THE IMMUNE SYSTEM HAS ALWAYS BEEN RELEVANT TO OUR COLLECTIVE health and well-being but as soon as you are at least semi-healthy, it is also pretty easy to ignore that side of life. But this all ended when suddenly a disease thing interrupted public and private life in a way that most people had never thought possible. A lot of terms and ideas from immunology were suddenly talked about regularly.

At the time this book was written the coronavirus pandemic was still raging, and there were a lot of open questions. A massive amount of research is being conducted around the world by countless scientists and we will learn a lot more in the coming years. In a way this is the best and the worst time to write a book about the immune system—the best time because more people might be interested in understanding what the hell is going on inside their bodies and how the body handles diseases. But the worst time because it would be so nice to write a comprehensive explanation of COVID-19, which is presently just impossible to do as there is still so much science going on.

But I think it still makes sense to talk a bit about it anyways. Luckily, all in all, immunologists have a solid grasp of the fundamentals of the coronavirus and what it does to us already. But first of all let us define what we are talking about.

Soon after the start of the pandemic, the virus, which has the horrible official name *Severe Acute Respiratory Syndrome Coronavirus 2,* was just called "coronavirus" by the public. Which is unfortunate and sort of wrong as coronaviruses are just a group of viruses, not a single species. But because of the rapid spread of the pandemic we missed our window to give this particular species of coronavirus a good, unique name. And while I complain a lot about scientists and the names they chose for things in this book, this time I really can't blame them as they were understandably pretty busy. When stressful things happen quickly we settle for whatever works at the time and that is OK.

So there are a lot of different coronavirus species, and they do a lot of different things. Mostly they infect the respiratory system of mammals, like bats and unfortunately humans.

Humans in particular are affected by a number of different coronaviruses. For example, around 15% of the cases of the common cold are caused by a coronavirus species. Coronaviruses have been around us for a long time, and many of the people reading this sentence already have antibodies against some of them flowing through their blood.

There were even dangerous coronavirus pandemics in the last few decades that you may even have heard about, like the *SARS coronavirus* (SARS is an abbreviation for "severe acute respiratory syndrome"). It is a respiratory disease caused by a coronavirus strain that also was found in bats in China in the early 2000s. It infected a few thousand, and killed a few hundred people, having a mortality rate close to 19%, which is pretty harsh.

A few years later, there was a second serious coronavirus outbreak. This time it originated in the Middle East and was called MERS, short for "Middle East respiratory syndrome." This one was even deadlier than SARS, and although it infected only around 2,500 people, it killed more than a third of them, with a horrible mortality rate of 34%. Both of these coronavirus species never got off the ground enough to become a real worldwide pandemic, which, if we look at their death rates, is something we can be truly grateful for.

Our collective coronavirus-related luck ran out at the end of 2019, when another coronavirus emerged. It is wildly more infectious than its predecessors but also far less deadly. Thanks to SARS and MERS, scientists had plenty of time to learn a lot about the mechanisms of dangerous coronavirus infections before the worldwide pandemic began.

Now, it is impossible to just describe with absolute certainty what happens during COVID-19, because it depends in large part on the patient. It has been widely reported that most people get no or only mild symptoms from an infection, while a minority has serious symptoms, often requiring hospitalization, and that an even smaller group dies. In diseases where symptoms vary that strongly from person to person, the reason is usually found within the individual's immune system and how it handles the infection. On top of that, the development of COVID-19 infections is pretty complex and new things are still being learned constantly. All of this makes COVID-19 hard to explain in detail, at least if this chapter wants to remain valid for a while. So we will stick with what we know, or, at most, what scientists are pretty confident about.

Some people infected by the coronavirus develop no symptoms at all, although they still seem to transmit the virus to others. Up to 80% of patients develop a mild disease, which still means pretty unpleasant symptoms for many. Mild, in this context, really only means that you don't require hospitalization. One of the first signs of an infection is often the loss of smell and sometimes even the sense of taste—which is a much bigger deal for the quality of life than most people realize until it hits them. In most people, the sense of taste and smell begins to return after some weeks. Although the virus has not been around long enough for us to know how long this sensory recovery will take.

Aside from that, most milder cases experience what could be called flu-like symptoms like fever, coughing, a sore throat, headache and body aches, and a general sense of exhaustion. A constant state of exhaustion, problems with concentration, and a reduced lung capacity also are symptoms that do not recede in some people, even months after they were infected.

But there remain a lot of open questions, especially about the long-term consequences for the people who experience them. At this point we simply don't know yet if the coronavirus pandemic will cause irreversible damage or not. In the cases of the more deadly SARS and MERS outbreaks, it took at least five years for the physical changes in patients' lungs to return to normal. What exactly does the coronavirus do and why is it so deadly to some?

The coronavirus targets a specific and very important receptor called *ACE2*. This receptor has a few vital jobs in your body, specifically regulating your blood pressure, which means you have a lot of cells in your body that carry it and can be infected. If you guessed that the epithelial cells in your nose and lungs have plenty of this receptor, you'd be right. From the perspective of a coronavirus, your lungs are miles and miles of free real estate.

But the ACE2 receptor also sits on cells in a variety of tissues and organs around your body. Your blood vessels and capillaries, your heart, your gut, and your kidneys. All of them have ACE2. As we learned before, your body's first response to a viral infection is chemical warfare, which basically does three main things: Interferons interfere with virus reproduction and slow it down, while other cytokines cause inflammation and alert immune cells.

Now, one thing that makes the coronavirus so dangerous is that it seems to be able to shut down (or strongly delay) the release of interferons, while the infected cells still release all the cytokines that cause inflammation and alert the immune system. So the virus is able to infect a lot of cells and spread quickly without being slowed down, while at the same time triggering widespread inflammation and activating immune cells that will themselves cause even more inflammation.*

And here is where things get dangerous for many people. The massive amounts of inflammation and active immune cells can cause serious damage to your lungs—if you remember, this is a region where your immune system usually tries to tread lightly because the tissue here is pretty sensitive. Without the interferon the virus still multiplies with very little resistance, while the inflammation is already causing damage.

As millions of epithelial cells die, now suddenly, the lungs' protective lining is gone and your alveoli, the tiny air sacs that do the actual breathing by exchanging gases between your insides and outsides, lie bare and can be damaged or even killed in the ongoing battle that ensues.

If it gets to this point, many critical patients will likely need to be mechanically ventilated, which is a fancy way of saying "sticking a tube into the lungs," and of course is a great way for bacteria to get a shortcut deep into your lungs where they find a pretty stressed-out immune system and a lot of tissue waiting to be colonized. This can get dramatic very quickly. If you are really unlucky you could also get a co-infection with more serious bacteria that can't believe their luck and enter the environment deep inside your

---

* A callback to what we learned before. One of the reasons why some people are able to deal better with the coronavirus than others is the genetic variability and the difference in MHC molecules or your toll-like receptors that lead to slightly different immune systems from person to person. Some immune systems are just better at dealing with the virus than others. And some are unfortunately really bad at dealing with it. So if you hear in the media that seemingly young and healthy people suffer from severe cases of COVID-19 and even die, this is one aspect of it. We never know what our individual immune system is great against until it is put to the test.

lungs. As the bacteria multiply the immune system has to react to the new threat and sends in the troops, more Macrophages and Neutrophils that do their job: Vomiting acid and causing more inflammation and damage.

Do you see the horrible pattern that emerges here? Stimulation causes activation, which causes more stimulation, which causes more activation, and so on. A horribly dangerously devious cycle with often deadly consequences. The massive amount of inflammation in the lungs can rip literal holes in the tissue and cause irreversible damage and scar tissue where your body hastily tries to heal. Even after surviving, many people may have a reduced lung capacity for the rest of their lives, which means trouble breathing and a reduced ability for physical activities.

In this context many people might also have heard about cytokine storms for the first time, which means a massive overreaction and overstimulation with all the signals your immune system is usually very careful to use just in the perfect amounts.

And we are still not done with the infection. And so there is even more bad news at this point: another critical body system can be affected by the typhoon of chemical screaming and overstimulation that is going on. In many serious COVID-19 cases, a coagulation cascade is triggered that leads to blood clotting, which means that the parts of your blood responsible for closing a wound can activate and begin to clot in your fine blood vessels, leading to a lack of oxygen supply in your organs. The body is now choked for oxygen from the inside while it also has a harder time breathing as the lungs fill with fluid. And of course, the clotting can cause a stroke or a heart attack, with all the known consequences.

For many people who had already been suffering from serious diseases, this is too much. Diabetes, heart disease, elevated blood pressure, and obesity are just some of the risk factors.*

On top of that, many older individuals just have weaker immune systems that did not have the most impressive interferon response to begin with, and

---

* One of the many reasons obesity is such an unhealthy thing is that fat tissue produces loads of inflammatory cytokines. So even on a good day, an obese person has a lot of inflammatory signals in their system. When infected by the coronavirus for example, their starting point is already worse, they are already more inflamed than they should be.

get overwhelmed much more easily by the coronavirus. This is the reason why most deaths occur in older individuals and the ones with preexisting conditions. But make no mistake, plenty of formerly healthy and young people die too. It is just a matter of bad luck and how your immune system handles all these challenges.

Let us end this chapter here. As I'm writing this sentence the world has begun vaccinating against COVID-19 and with a bit of luck, as you are reading this sentence, we are all returning to a world that feels normal again. In any case, the coronavirus pandemic was a stark reminder of why your immune system is so incredibly important and why more people would benefit from understanding it better.

Immune System:
An Overview

Pathogens
Bacteria, Viruses etc.

Neutrophil
Kills, communicates, causes inflammation

Macrophage
Communicates, activates other cells, kills enemies, causes inflammation

Complement
Mark and crippling enemies, activate and guide immune cells

Dendritic Cell
Identifies enemies, activates other cells

Infected Cell

Natural Killer Cell
Communicates, kills infected/cancer cells

Monocyte
Becomes macrophage, identifies and kills

Eosinophil
Causes inflammation, battles parasites, activates other cells

Basophil
Causes inflammation, battles parasites, activates other cells

Mast Cell
Causes inflammation, communicates, activates other cells

Parasitic Worms

Innate Immunity

Attack and Kill
Activate
Communicate
Produce Antibodies
Morph

Virgin Killer T Cell
Standby mode,
kills infected/cancer
cells

Memory Killer T Cells
Remember enemies,
kill infected/cancer
cells

Infected Cell

Killer T Cell
Kills infected/cancer
cells

Memory Helper T Cells
Remember enemies,
communicate,
activate

Virgin Helper T Cell
Standby mode,
activates other cells

Helper T Cell
Communicates,
activates other cells

Virgin B Cell
Standby mode,
activates other cells

Long-Lived Plasma Cell
Produces antibodies

Plasma Cell
Produces antibodies,
activates other cells

B Cell
Produces antibodies,
activates other cells

Antibodies
Mark and disable enemies,
activate complement

Memory B Cell
Remembers enemies,
produces antibodies

Adaptive Immunity

A Final Word

As with every good journey, arriving somewhere is as important as leaving in the first place. We saw a great many things and many complex and intertwined systems. We got to know all your surfaces, inside and outside, and their intricate defense networks. We met your soldiers, from black rhinos that are calm most of the time to crazy monkeys with machine guns.

We observed how your immune system jumps into gear when your body is breached and wounded, how multiple layers of complexity work together to organize exactly the correct type of defense over distances that are enormously large for your tiny cells. We visited the largest library in the universe and the deadliest university that you carry with you without even thinking about it.

We witnessed a sneaky attack on your most inner self by an army of viruses that was as effective as it was cruel and uncaring. We explored how your immune system remembers its battles and how we as humans can assist it with that. We took a look at what happens when your immune system fails or when it overcommits and becomes the source of disease and damage. And while we did dive pretty deep at times, there are so many more amazing places and systems we did not have time to visit. But if you made it to this page, you have gotten a real roundtrip through your own body and some of the most important things you probably never thought about.

An annoying thing about the immune system is that you need to understand multiple things at the same time before the whole system begins to make sense, and before its true beauty reveals itself. If you understand Macrophages and MHC molecules and cytokines and T Cell receptors and the lymphatic system and antibodies, then they all combine into an amazingly elegant system that makes so much sense and is pretty stunning.

But getting started is extremely hard because the immune system seems to be designed to be opaque and hard to grasp. I complained a bunch about the language of immunology and while that was hopefully mildly amusing to you, in reality, it was not so much for me. To research this book I had to read textbooks and academic papers with the speed of a first grader, just so I could keep up with what they tried to say. I can't imagine a field that would profit more from cleaning up its language and making an effort to become more palatable to the general public. Because in the end, immunology is truly one of the coolest topics ever.

Science offers such a diverse array of topics that you can immerse yourself in. And in popular culture it is often the seemingly large topics and fields that are the most beloved. Space, for example, with its huge distances and black holes and gigantic stars, is an easy sell for documentaries and popular science books. But while space is nice and all, it has nothing over biology. Stars are dead clumps of burning plasma, and even the most complex and interesting one can't compete with the wonder and complexity of the simplest bacteria trying to escape a Macrophage.

The immune system is not as pleasing, not as accommodating as other fields of popular science. It asks a lot of you up front. A certain investment of time and pain is necessary to get to the point where you can really appreciate it. And in a time where the expectation is that information has to be pleasing and easy to digest this feels like a lot to ask. Despite these challenges, the immune system is one of the *best* topics to learn about because of the fact that it is so complex and made up of so many layers that all interact in such ingenious ways—it is like a window into the universe itself. A window into the complexity that surrounds you and that you are a part of. You are incredibly lucky to be alive and to have a body that you can call your own. Or at least I feel that way.

So I would argue that it is worth the investment because the payoff is so amazing and I hope if you read this far, you feel the same. Once you reach the mountaintop and get a sort of clear picture of the immune system, the view is like no other. You get a taste of what it means to stay alive in a world that is a struggle between different forces that do not care about how you feel about them.

All of this beautiful complexity carries a hint of sadness. It stings a bit to know that life is too short and too busy to truly learn about all the layers that make up reality. But hey, in the end there is nothing we can do about that. What we can do is to take up the challenge from time to time and put in the effort to get a glimpse into something so much larger than us.

Even if we will never get to the bottom of it.

Sources

Publishing printed things is weird because you need to finish so early before actual publication. So in order to save time and make the lives of the printers easier, a detailed bibliography of the papers and books used for the research of the book can be found online at https://kurzgesagt.org/immune-book-sources/.

Acknowledgments

THIS BOOK WOULD NOT EXIST WITHOUT THE GENEROUS HELP OF EXPERTS who took time for me from their busy schedules doing actual science and stuff. They patiently answered my many questions, steered me in the right direction when I got lost in the research, told incredible stories about the immune system and its adversaries, and were just incredibly fun to talk to. All while being occupied with making the world better during a global pandemic that did not make anybody's life easier.

So a huge thank you to Dr. James Gurney, who gave extensive feedback, did a lot of fact checking, and told exciting stories from the world of microbes and viruses. Book bro fist to Professor Thomas Brocker, the director of the Munich Institute for Immunology, for jumping on many video calls to answer multitudes of often weird questions about details of immunology. And transatlantic high five to Professor Maristela Martins de Camargo of the University of São Paulo for the many amazing and mysterious stories about all the crazy things our immune cells do!

I would never have dared to publish a book about such a complicated topic without your help and I remain incredibly grateful for your time and enthusiasm. On top of that, it was such a genuine blast learning from all of you and I hope after the pandemic is over we can touch glasses at some point!

Then I want to thank my friends Cathi Ziegler, John Green, Matt Caplan, CGP Grey, Lizzy Steib, Tim Urban, Philip Laibacher, and Vicky Dettmer, all of whom read the whole book at various stages of completion, some of them multiple times. Thank you all for your detailed feedback and conversations about the right tone, letting me know if jokes landed or if the explanations worked. Thank you for being brutally honest with me when necessary and for being encouraging when I was down and didn't believe this book was possible to finish. It's a huge ask for a friend to read a whole book and then give feedback, especially if it's not finished yet, so I'm extremely grateful that you took the time. Thank you so much.

Thank you to Philip Laibacher, the first employee and Creative Director of Kurzgesagt–In a Nutshell, for creating the beautiful illustrations in the book and the amazing cover. And thank you for sacrificing a part of your Christmas vacation so everything could be done in time.

Of course I also owe a huge thanks to my agent Seth Fishman from the Gernert Company for calming me down when I felt a bit panicky about writing my first book and for getting this whole thing started. To my editor, Ben Greenberg from Random House, for believing in this project, editing the early drafts and pushing them in the right direction, and just for being a calm presence in this whole process. Thank you both for not laughing at me when I confidently said, like an idiot, that I'd finish this book in three months. To Kaeli Subberwal, Rebecca Gardner, and Jack Gernert for patiently dealing with me as I was a cliché author type who never answered his emails. A big thank you to all people at the Gernert Company and Random House for dealing with me and being great at their jobs and being so positive and making this book possible.

I also want to thank my whole team at Kurzgesagt–In a Nutshell. I sorta just took a prolonged leave to write a book that was dear to my heart, and my team had my back and kept the channel and the company running. Sorry for being bad at communicating at times—I appreciate all of you and the work you do.

A big thanks to all the viewers and fans of Kurzgesagt. Like, I don't know most of you personally and I never know what to say when someone tells me to my face that the work my team and I do means something to them. But here in the safety of a printed page: Thank you for liking the stuff I write and thank you for supporting it. It means the world.

And if you read this book and got this far: There are so many other things you could have read but you read this thing. So thank you.

Index

About the Author

PHILIPP DETTMER is the founder and head writer of Kurzgesagt, one of the largest science channels on Youtube with over fourteen million subscribers and one billion views. After dropping out of high school at age fifteen, Dettmer met a remarkable teacher who inspired in him a passion for learning and understanding the world. He went on to study history and information design with a focus on infographics. Dettmer started Kurzgesagt as a passion project to explain complicated ideas from a holistic perspective. When the channel took off, Dettmer dedicated himself full-time to making difficult ideas engaging and accessible.

philippdettmer.com
YouTube: Kurzgesagt—In a Nutshell

About the Type

This book was set in Scala, a typeface designed by Martin Majoor in 1991. It was originally designed for a music company in the Netherlands and then was published by the international type house FSI FontShop. Its distinctive extended serifs add to the articulation of the letterforms to make it a very readable typeface.